Human and Nonhuman Bone Identification

A Concise Field Guide

Human and Nonhuman Bone Identification

A Concise Field Guide

Diane L. France

CRC Press
Taylor & Francis Group
Boca Raton London New York

CRC Press is an imprint of the
Taylor & Francis Group, an **informa** business

CRC Press
Taylor & Francis Group
6000 Broken Sound Parkway NW, Suite 300
Boca Raton, FL 33487-2742

© 2011 by Taylor and Francis Group, LLC
CRC Press is an imprint of Taylor & Francis Group, an Informa business

No claim to original U.S. Government works

Printed in the United States of America on acid-free paper
10 9 8 7 6 5 4 3 2 1

International Standard Book Number: 978-1-4398-2039-1 (Paperback)

This book contains information obtained from authentic and highly regarded sources. Reasonable efforts have been made to publish reliable data and information, but the author and publisher cannot assume responsibility for the validity of all materials or the consequences of their use. The authors and publishers have attempted to trace the copyright holders of all material reproduced in this publication and apologize to copyright holders if permission to publish in this form has not been obtained. If any copyright material has not been acknowledged please write and let us know so we may rectify in any future reprint.

Except as permitted under U.S. Copyright Law, no part of this book may be reprinted, reproduced, transmitted, or utilized in any form by any electronic, mechanical, or other means, now known or hereafter invented, including photocopying, microfilming, and recording, or in any information storage or retrieval system, without written permission from the publishers.

For permission to photocopy or use material electronically from this work, please access www.copyright.com (http://www.copyright.com/) or contact the Copyright Clearance Center, Inc. (CCC), 222 Rosewood Drive, Danvers, MA 01923, 978-750-8400. CCC is a not-for-profit organization that provides licenses and registration for a variety of users. For organizations that have been granted a photocopy license by the CCC, a separate system of payment has been arranged.

Trademark Notice: Product or corporate names may be trademarks or registered trademarks, and are used only for identification and explanation without intent to infringe.

Visit the Taylor & Francis Web site at
http://www.taylorandfrancis.com

and the CRC Press Web site at
http://www.crcpress.com

Contents

Preface vii

Acknowledgments ix

Part I: General Osteology

Introduction 3

Part II: Major Bones of the Bodies of Different Animals

Cranium 47

Mandible 67

Scapula 87

Humerus 107

Radius 127

Ulna 151

Metacarpals and Forelimbs 169

Pelvic Girdle 187

Femur 207

Tibia 227

Fibula 247

Metatarsals and Hindlimbs 257

Index 273

Preface

This book is the field version of *Human and Nonhuman Bone Identification: A Color Atlas*, published in 2009. It presents the major skeletal elements from the same species as the previous book, but in a smaller format more easily taken into the field. This field guide does not include all of the photographs from the previous book, nor does it include the section that illustrates all of the bones from each animal together, but it does present a preview of the bones of a bird that will be featured more thoroughly in a future volume that will include skeletons of human subadults, birds, amphibians, reptiles, and fish.

This field guide, like the original book, is intended to give law enforcement, medicolegal death investigators, forensic anthropologists, and even the general public a guide with photographs and other information necessary for comparing adult human bones to adult mammals. This book is presented in two major sections:

I. General Osteology, including the major features of bone, growth and development, and general comparisons of quadrupedal mammals to human bones. This section also includes an introduction to bird skeletal anatomy and some suggestions on how to clean and preserve bone.

II. Major Bones of the Bodies of Different Animals (grouped by bone).

This book is not intended to give all of the answers, and it is certainly not intended as a replacement for experienced professionals! It is intended as a rough guide to help determine human from nonhuman mammals. Not all of the bones of the body are represented in this book. It is logical, and it has been my experience that the bones represented in this book are those most often discovered.

There is natural variation among individuals within species (including variation between the sexes) and this variation cannot be fully displayed within this book. Be aware that this may lead to misdiagnosis!

This book discusses the gross morphological differences between the species. It does not address the biochemical or microscopic differences between humans and nonhumans. If a bone or bone fragment cannot be identified with help from this book, be aware that other methods exist to identify whether or not a bone is human.

IF YOU ARE UNSURE OF YOUR DIAGNOSIS, CONSULT A PROFESSIONAL!

A note about the language used in this book

Although it is typical to hear someone ask whether a bone is "human or animal," it is true that humans ARE animals, and a more appropriate and technically correct way to ask the question is: "Is it human or nonhuman?"

Also, because this book is intended for the researcher who is trying to distinguish between humans and nonhumans, although nonhuman skeletons have different names for some of the skeletal and dental elements, the human equivalent term is used.

Acknowledgments

Authors of many books say that the book could not have been completed except with the help of others, and that statement is no truer than with this book. If I had to single out one person without whose help this book would not have been possible, that person would be Dr. Linda Gordon, Collections Manager, Division of Mammals at the Smithsonian Institution in Washington, DC. Although Dr. Gordon had as much work to do and as many deadlines as any person can have, she always offered me her time, expertise, and access to the collections, and for that I am most grateful. I owe her a great deal.

For this guide, another individual was extremely generous and helpful. Jay Villemarette, the owner of Skulls Unlimited International, Inc. in Oklahoma City, Oklahoma and his family and associates were welcoming and kind, and allowed me access to his vast collections. Jay deserves much praise and support for the important work that he does to preserve skeletal material and to educate the public about bones and the preservation of species.

My husband, Art Abplanalp is always supportive when I am writing and in every other aspect of our lives together. I am lucky to be with him.

Shane Walker, the manager of my small business, France Casting, took over most of the responsibilities of the business so that I could focus on this book. He is doing a great job and he deserves my thanks.

Becky Masterman and Jill Jurgensen from CRC Press/Taylor & Francis Group have been, as always, terrific. They are always quick to answer emails and I have thoroughly enjoyed working with them once again.

With all of this help and support, this book should be perfect. To the extent that it is not, it is my fault.

Part I

General Osteology

Introduction

What Is Bone? 10
    Bone Morphology 11
    Anatomical Terminology 16
    Features of Bone 17
    Comparisons of the Skeletons of Quadrupeds and a Biped 17
        The Vertebral Column and Thorax (chest) Area 18
        The Cranium 21
        The Pelvis..22
        The Limbs 23
        Growth and Development 25
        Dental Growth, Development, and Eruption 30

Special Notes about Bird Skeletons 31
    Features of a Bird Skull 34
    Postcranial Bird Skeletal Elements 36

Cleaning and Storing Skeletal Elements 41
    Dermestid Beetles 41
    Water Cleaning Methods..42
    Drying and Storing Skeletal Elements 42
    Cautionary Notes 43
    Preserving Skeletal Elements 43

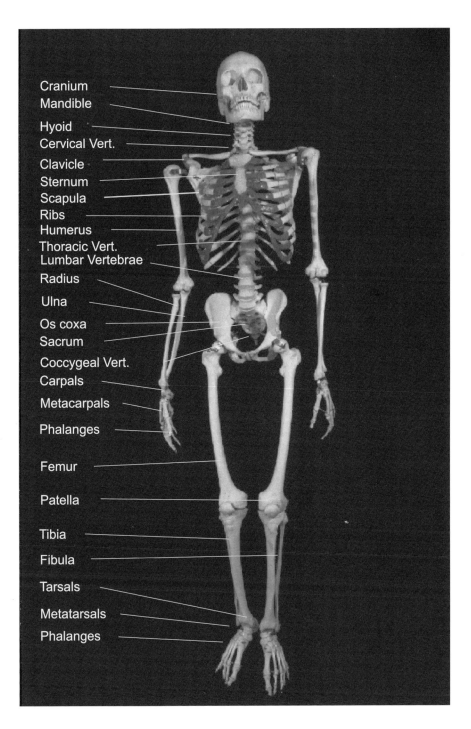

Figure 1.1 Human skeleton.

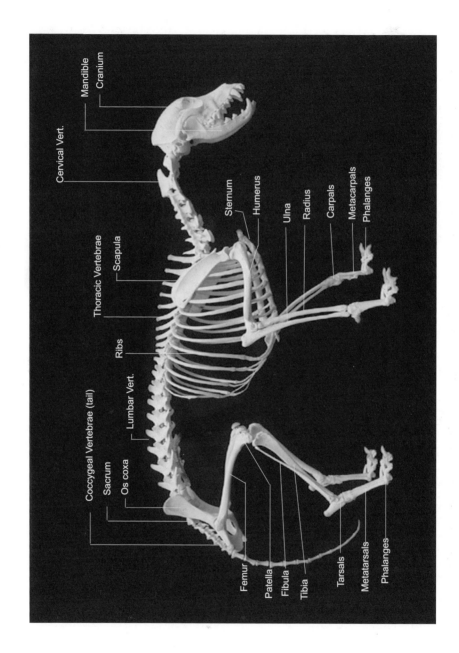

Figure 1.2 Dog skeleton.

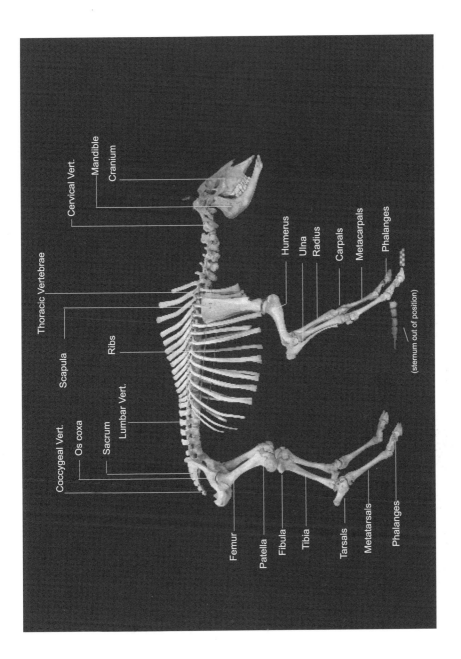

Figure 1.3 Bison skeleton.

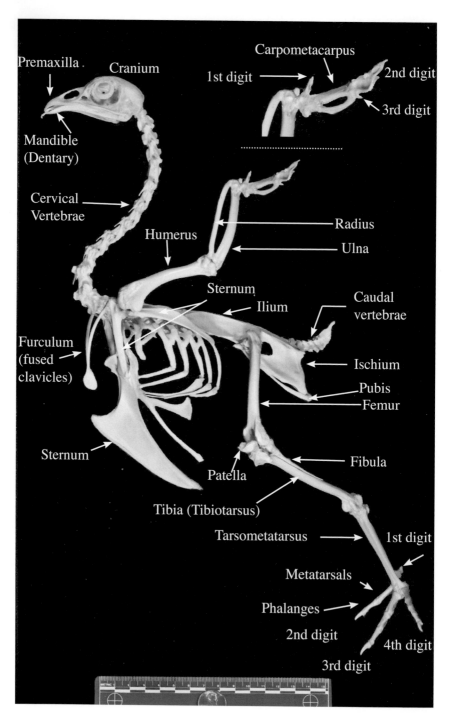

Figure 1.4 Bird skeleton.

Introduction

Before diagnosing whether or not a bone is human (indeed, at the start of any forensic investigation involving suspected skeletal remains), the first step is to determine whether or not the object in question is actually bone. Many organic and inorganic materials can mimic bone (see Figure 1.5). This can be even more confusing because bone can take on the color of its environment (bone can be darker when in dark soil, red in red soil, greenish when exposed to copper, and can be bleached white when exposed to the sun, wind, and water) (see Figure 1.6).

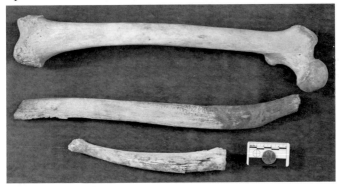

Figure 1.5 Human femur (top), wood (middle), very weathered bone (bottom).

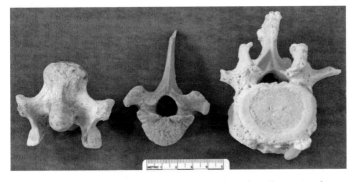

Figure 1.6 Bleached white vertebra that has been in the elements (left), vertebra that was discovered in dark soil (middle), vertebra that has been cleaned and preserved (right).

While the color of bone is not as important as other considerations when diagnosing species, it is very important in determining the *taphonomic* influences at work. *Taphonomy* is defined as anything that happens to a body after death. This includes the decomposition environment and patterns (climate, water and insects, for example, and even the temperature of the laboratory in which the remains are stored). The postmortem (after death) history of the remains is sometimes one of the most important clues in solving a forensic case, and should never be dismissed when collecting evidence (including remains).

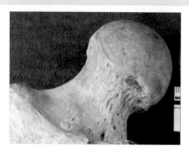

Figure 1.7 Human femoral neck. Note external texture differences between the femoral head and the neck.

In addition to the general gross morphology (shape) of the skeletal element, the external and internal texture of the bone is vital to diagnosing the bone and the species (see Figure 1.7 for an example of texture on bone). The basis for understanding why this is important involves understanding the different components of bone.

What Is Bone?

Bone is comprised of both organic and inorganic components. That is, bone is not entirely mineral; there is a soft tissue component as well. The mineral component is a compound of calcium and phosphates called hydroxyapatite that is formed in and around an organic matrix containing collagen. Collagen is similar in consistency to very thick, relatively hard gelatin (similar to Jello© made with much less water than usual). In living bone and in bone that is still relatively fresh after death, the collagen component is significant, but as the body and bone decompose, this organic collagen component usually decays before the mineral component is significantly affected. There are exceptions to this rule for example, if, the bone is exposed to chemicals that dissolve the mineral component and leave the organic component (most people will remember grade school biology and soaking chicken bones in vinegar to remove the minerals and leave a rubbery, soft material that looks like a chicken bone). Remembering that there is an organic as well as an inorganic component to bone also helps to explain the way in which bone develops and the way it reacts to various stresses (fractures, cuts, disease, etc.).

Bone Morphology

Bone is a living, dynamic tissue that responds to its environment. To a large extent, the form of a bone is determined by its function and the function is determined by its form. For example, humans use our forelimbs largely for manipulating and carrying objects while a cow uses its forelimbs for locomotion and to support the cranial half of its body. It makes sense, therefore, that a cow forelimb will be more massive and have a narrower range of motion (for stability) than will the human forelimb. If an investigator understands and uses this basic principle, it will not be necessary to memorize the form of each bone of each species to diagnose whether or not the bone is human!

It is interesting, however, that the individual bones of human and nonhuman mammals (the concentration of this book) are similar enough in morphology that it is relatively easy to determine whether the bone is an ulna or a femur (Figures 1.8 and 1.9). If the investigator is able to determine which bone of the body he is holding, it is easier to determine if the bone is human.

Figure 1.8 Ulna of human (left) and deer (right).

Figure 1.9 Femur of human (left) and moose (right).

In determining whether or not a bone is human, it is important to distinguish between an area of bone-to-bone articulation, an area of muscle attachment (origin and/or insertion) and an area of relatively smooth bone that is neither an area of articulation nor an area of muscle attachment (Figure 1.10). In healthy bone, the area of articulation between two bones designed to move against each other* will have a smooth surface. This surface will be separated from the articular surface of the other bone by a layer of cartilage able to withstand normal movement, and is sometimes filled with a slippery lubricant called synovial fluid (somewhat like egg whites and, in fact, "syn" means "together" and "ovia" means "egg"). If the cartilage or articular surface is damaged, the joint surfaces may break down causing degeneration and perhaps areas of eburnation (polishing) caused by bones polishing each other by rubbing action (Figure 1.11). This and other pathological conditions may confuse the diagnosis of species, and if a pathological condition is suspected, the bone should be taken to an expert for diagnosis.

---

*There are other types of joints between bones in which the motion is limited or essentially absent (for example, the sutures of the cranium). These articular surfaces are not smooth.

Introduction

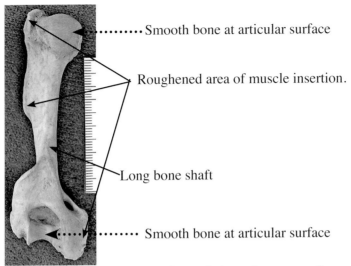

Figure 1.10 Articulation and muscle insertion areas of a common long bone.

Figure 1.11 Degenerative process in knee joint with breakdown of the articular surface and eburnation (polishing).

The area of origin or insertion of a muscle (or muscle tendon) or ligament on bone is rough and often raised (although not all rough areas are locations of muscle attachment). Generally, the larger and more powerful a muscle is, the more area of bone it needs on which to anchor itself.

Notice, for example, the large crest on the back of the cranium on the moose in Figure 1.12a and compare that to the smoother corresponding area on the human (Figure 1.12b). The neck muscles in the moose must work against gravity to hold up a very large head, while the head of the human is balanced on top of the spinal column and does not require large muscles to hold up the head.

Figures 1.12 Moose cranium (left) showing large area for muscle insertion at arrow and human cranium (right) showing smaller area for muscle insertion.

After identifying the bone in question (femur, humerus, etc.) and identifying the areas of articulation and muscle insertion on the bone, one can determine whether the bone is from a quadruped and whether or not it is from a mature individual. Figures 1.10, 1.13, and 1.14 show common features and terminology used in osteological analysis.

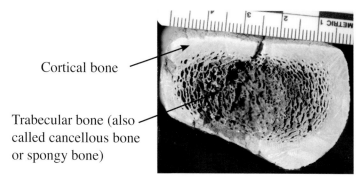

Cortical bone

Trabecular bone (also called cancellous bone or spongy bone)

Figure 1.13 Cross section of typical long bone.

Introduction

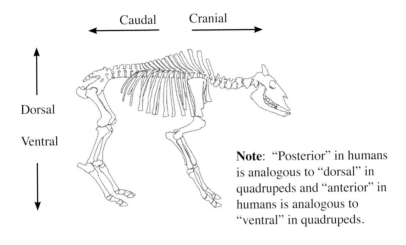

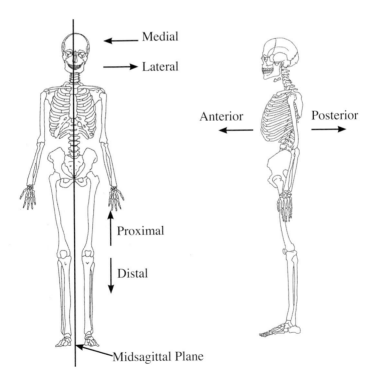

Figure 1.14 Planes of the body.

ANATOMICAL TERMINOLOGY

Anterior: in front (analogous to ventral in humans)
Appendicular: the skeleton of the limbs
Axial: the skeleton of the head and trunk
Caudal: toward the tail
Coronal plane: parallel to the coronal suture
Cranial: toward the head
Distal: away from the trunk of the body along a limb
Dorsal: in back (analogous to posterior in humans)
External: outside of
Inferior: lower
Internal: inside of
Lateral: perpendicularly away from midsagittal plane
Longitudinal: coursing or placed lengthwise
Medial: perpendicularly toward the midsagittal plane
Midsagittal plane: in a line defined by the sagittal suture of the cranium
Posterior: behind, to the back (analogous to dorsal in nonhumans)
Pronation: rotation of the hand and forearm so that the palm faces dorsally or toward the body
Proximal: toward the trunk of the body along a limb
Sagittal section: any section of the body parallel to the sagittal suture of the cranium
Superficial: near the surface
Superior: above, top
Supination: turning the palm of the hand anteriorly
Transverse: any crosswise section
Ventral: in front (analogous to anterior in humans)
Vertex: top, highest point

In mouth:

Buccal: toward the cheek
Distal: at greatest distance from the anterior midline of the mouth
Labial: toward the lips
Lingual: toward the tongue
Mesial: toward the anterior midline of the mouth
Occlusal: the chewing surface of the teeth

Introduction

FEATURES OF BONE

Feature (plural)   **Definition**

Cavity (cavities): an open area
Condyle (condyles): rounded process at the point of articulation
Crest (crests): a projecting ridge
Diaphysis (diaphyses): the shaft of the bone (and primary growth center)
Epiphysis (epiphyses): secondary center of bone growth attached to the diaphysis and usually consolidated with it by bone
Fontanelle (fontanelles): membranous space between cranial bones in fetal life and infancy
Foramen (foramina): a hole or opening
Fossa (fossae): a pit, depression or cavity
Meatus (meatuses): a canal
Process (processes): any outgrowth or prominence of bone (also, in this book, called a projection
Sinus (sinuses): bone cavity lined with mucus membrane
Suture (sutures): areas of articulation between cranial bones
Torus (tori): an elevation or prominence
Tubercle (tubercles): a small, knob-like projection on bone
Tuberosity (tuberosities): a large, rough eminence or projection on bone

COMPARISONS OF THE SKELETONS OF QUADRUPEDS AND A BIPED

Hints:

More sculpted bones are usually nonhuman, even in immature bones.

More sculpted articular surfaces have decreased range of motion while less sculpted articular surfaces may have greater range of motion.

Quadruped: an animal that habitually walks on four limbs
Biped: an animal that habitually walks on two limbs.

This book is intended to be a guide to the differentiation between humans and nonhuman quadrupeds. Quadrupeds (or those animals who habitually walk on four limbs) and bipeds (humans who walk on two legs) are not the only two categories of locomotion, but they are the two categories pertinent to this book (a short section on bird bones and how they differ from mammals is included at the end of this section). For example, brachiators (those primates who habitually swing from branch to branch in trees) have significantly different skeletal morphology because of this greatly different locomotion pattern, and they will not be covered in this book.

Among the quadrupeds covered in this book, the differences in locomotor patterns are reflected in the morphology of the skeleton (particularly the postcranial skeleton). The skeleton of a deer reflects the need for an animal of moderate size to move quickly. Elk need to move quickly, but they are larger animals than deer, and the skeleton reflects that size difference (the bones are larger and more massive). The buffalo and cow are massive animals that do not move as quickly, so their skeletons must support much more weight without the great speed. Horses are large, fast animals, and their skeletons are interesting in that they are significantly different from any other animal studied in this book. Sheep and goats are relatively short animals with significant weight for their height and a moderate need for quickness. Dogs, and particularly cats, are fast runners that do not carry much weight. Beavers show skeletal modifications near the tail that reflect large muscle insertions to control the large tail. Badgers are digging animals and their forearms demonstrate that pattern.

The Vertebral Column and Thorax (Chest) Area

The vertebral column is divided into five sections: cervical (usually 7 in number), thoracic (usually 12), lumbar (usually 4 to 6), sacral (usually 4 to 6 but fused in the adult to form the sacrum) and coccygeal (varies in number according to whether or not the species has a tail). Humans and nonhumans have about the same number of vertebrae (even giraffes have only 7 cervical or neck vertebrae! see Figure 1.15), but the shape of the vertebral column and of the individual vertebral bodies differs. The vertebral column in a typical quadruped has a single gradual curve from the neck to the pelvic girdle (somewhat like a cantilever bridge),

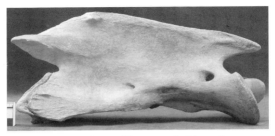

Figure 1.15 Second cervical vertebra of human (left) and giraffe (right) placed in the same plane. Notice the size of the size of the scale in each photograph (smallest ticks are millimeters in each).

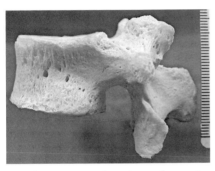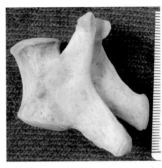

Figure 1.16 Wedge-shaped vertebra of human (left) and cylindrical vertebra of mountain lion (right).

while the human has an S-shaped column. The difference in vertebral column shape is reflected in the morphology of the vertebrae as well. The quadruped typically has longer, more cylindrical vertebral bodies than does the human, and the vertebral bodies are more similar in length from the neck region to the pelvis. Humans have more wedged-shape vertebrae (Figure 1.16), and the bodies of the vertebrae are gradually larger from the neck to the pelvis (each vertebra carries more weight than the vertebra above it, so the bodies are larger as one progresses down the vertebral column).

The spinous process of a vertebra (in all species) is that projection on the dorsal or posterior aspect of the vertebra (dorsal in a quadruped is analogous to posterior in humans). The spinous process is the area of

Figure 1.17 Spinous process of thoracic vertebra in bison (above) and human (right).

 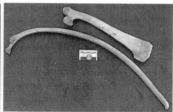

Figure 1.18 Typical ribs of human (left) and cow (right).

Figure 1.19 Superior (left) and inferior (right) human clavicle.

muscle insertion along the spine, and are different between large quadrupeds (cows, horses, etc.) and humans or small quadrupeds. This is the general area of insertion of the neck muscles responsible for holding the head up against gravity. Note that in the large quadrupeds the spinous processes are very large relative to the size of the vertebral body (see Figure 1.17).

Note that the thorax (chest cavity including ribs) is deep and narrow in quadrupeds and shallow and broad in humans, which brings the center of gravity of humans closer to the vertebral column. This, naturally, changes the shape of the ribs, making the ribs straighter in quadrupeds and more curved in humans (see Figure 1.18).

The clavicle maintains the distance between the sternum and scapula and provides support for the shoulder girdle. It is present in humans and in some other mammals in which the forelimbs are used for manipula-

 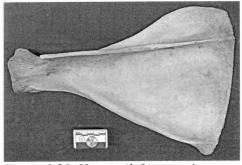

Figure 1.20 Human (left) posterior scapula and moose (right) dorsal scapula.

tion (such as the beaver), and in birds. It is vestigial or absent in many mammals and is therefore of limited use in species identification, other than to alert the investigator that the bone could be human. The clavicle of the human is shown in Figure 1.19. In both photographs, the sternal articular surface is on the left and the scapular articular surface is on the right.

The scapula is elongated in most nonhuman mammals with the glenoid fossa (the point of articulation with the humerus) at the end of the long axis. In humans the scapula is more triangular in shape with the glenoid fossa along the most lateral surface (see Figure 1.20).

The Cranium

As stated above, the area of the occipital region of large quadrupeds is modified for the attachment of large neck muscles devoted to counteracting the effects of gravity on a large skull. The nasal region of many quadrupeds is long and narrow. The increased sense of smell is reflected in this long nose in many animals, though in some animals the length of the face is a reflection of the morphology of the dentition. For example, the canids have as keen a sense of smell as carnivores while the horse has a dental complex that reflects its vegetarian diet. The foramen magnum of a typical quadruped is located more posteriorly (which make sense because the skull is in front of the spinal column). The foramen magnum in a biped is more centrally located under the cranium, which helps in balancing the cranium on the vertebral column (Figure 1.21). The mastoid process (see Figure 1.22) is the point of insertion

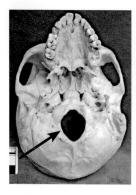
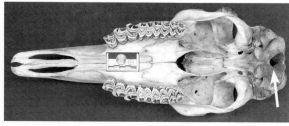

Figure 1.21 Foramen magnum in the human (left) and moose (above).

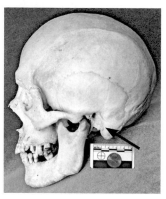

Figure 1.22 Mastoid process in human (left, at arrow) and corresponding area in wolf (right).

of the sternocleidomastoid muscle which originates on the clavicle and sternum, and is responsible for maintaining the balance of the skull on top of the vertebral column and for turning the head. The mastoid process is very small in quadrupeds, as there is little need to bring the cranium from a dorsal to a ventral position.

The Pelvis

Because of the changes in the ilium, the center of gravity in quadrupeds is different from that in the biped. The pelvic girdle (os coxae and sacrum) in quadrupeds is long and narrow and reflects the function of the leg muscles that attach to the pelvis. The lower limbs in the large quadrupeds move anteroposteriorly with very little lateral motion, so the strength of the muscles in the leg that make this movement possible are benefited by a long pelvis (which acts as a long lever arm). The

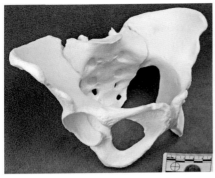
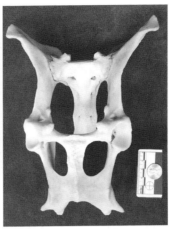

Figure 1.23 Human pelvic girdle (above), pig pelvic girdle (right).

pelvic girdle in humans has become shorter and wider, reflecting the different locomotion patterns (balancing the weight over each leg independently as forward movement occurs) as well as the difference in the support of the abdominal contents and the need for a large pelvic outlet for childbirth in females (Figure 1.23).

The Limbs

In general, the forelimbs and hind limbs of quadrupeds are of roughly equal length, while in humans the hind limbs are considerably longer than are the forelimbs. The forelimbs of most quadrupeds carry somewhat more weight than do the hind limbs, as the center of gravity is usually closer to the forelimbs.

Certain bones of the forelimbs and hind limbs of many quadrupeds are modified to increase the power to the legs. The concept of lengthening certain bones (and therefore muscle attachment areas) to increase the power of the muscle is easy to understand if we liken it to jacking up a car to change a tire. If you are trying to use a jack to lift a car, would you use more energy if you used a short handle or a long one? Naturally a long handle would use less of your energy and it would move through a greater distance to get the car off of the ground. The biomechanics of a long lever arm in animal locomotion works the same way. The animal uses less energy to move what is on the end of the long lever arm. In addition (and this is particularly important in animals that run at high speed), that lever (or leg) moves through a greater range of motion than does a leg that is shorter or that has a shorter lever arm.

The radius, ulna, tibia, and fibula allow rotary (pronation and supination) motion in humans and in most smaller mammals. In humans, the radius and ulna are roughly equal in size and allow great flexibility in pronation and supination (Figure 1.24). The tibia and fibula in humans still allow a little rotary motion in the foot (although it is greatly reduced as compared to other primates). In many small quadrupeds the tibia and fibula are still separate bones and allow some rotary motion, but in many of the large quadrupeds the fibula is greatly reduced resulting in no rotary motion of the foot. Likewise, in many of the large quadrupeds the radius and ulna fuse in the adult resulting in no rotary motion of the forelimb.

In general the articular surface of the limbs of quadrupeds such as dogs, cats, and horses are more sculpted than those of primates (and higher primates and humans in particular). Observe the articular surface of the distal femur of a moose and compare it to the human distal femur (Figure 1.25).

The hands and feet reflect different lifestyles in mammals. Most mammals have five fingers and five toes, but the larger quadrupeds have reduced fingers and toes — the mammals (such as cows and sheep)

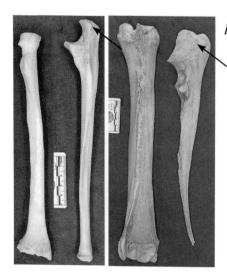

Figure 1.24 Human radius and ulna (left) and moose radius and ulna (right). Arrows point to olecranon process of the ulna, insertion point for muscles that extend the leg. Note the increased area of bone devoted to muscle insertion in a large quadruped and observe that the two bones in humans are roughly the same size, allowing pronation and supination.

Introduction

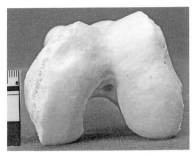
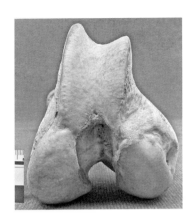

Figures 1.25 Human distal femur (above) and moose distal femur (right).

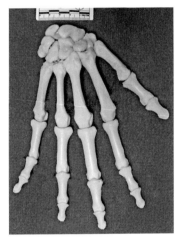
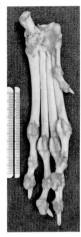
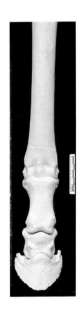

Figures 1.26 Hand of human (left), paw of wolf (middle) and hoof of horse (right)

have two digits while the horse has one. These mammals often have rudimentary digits higher up the foot. The dog and cat have four pads that touch the ground, but also have a "dew claw" higher on the forelimb and hind limb (Figure 1.26).

Growth and Development

One of the most confusing aspects of determining whether or not a bone is human is when one is trying to diagnose an immature bone. Very young bones (fetal or, depending upon the species, varying lengths of time after birth) are not as "sculpted" as are adult bones. Human bones, for the most part remain less "sculpted" throughout the life of the individual.

As was mentioned earlier, bone is comprised of both an organic and an inorganic component. Most bones of the mammalian body are first formed as a cartilage matrix, although some bones, such as many bones of the cranium, develop first from a different kind of soft tissue (membrane). The initial cartilage matrix grows in the fetus, and at some point in the bone's development it begins to be transformed into bone. When this occurs depends not only on the species but also the individual bone within the body, as different bones will develop at different rates and ages throughout the body. Experts can take advantage of this fact in diagnosing the age of an individual.

At the very earliest stages, the centers of bone growth start as a single bone cell, and for a time are indistinguishable from other centers of bone growth of the same size in the body. Their location in the body can, of course, be determined if the body is intact, and this can give valuable information about the age of the individual, but if the amorphous centers are discovered dry and out of context, they are often impossible to differentiate.

In the cartilage model, osteogenic (bone-forming) cells overtake the cartilage cells and replace them with bone a bone cell at a time. Often a single bone goes through this process at different parts of the bone at different times. The first area of the matrix to be replaced in a long bone is usually the approximate center of the shaft called the *diaphysis* (plural: *diaphyses*). Secondary centers of bone growth can occur at the ends of the bone and are each called an *epiphysis* (plural: *epiphyses*), and usually begin bone growth later than does the diaphysis. Until the bone ceases growth, there is a cartilage matrix between these centers. there is a unique surface at the ends of these growth plates (see Figures 1.27 and 1.28).

A bone may have several secondary centers of growth (Figure 1.29). In bone that starts with the cartilage matrix, each secondary center will grow and develop and eventually fuse into the growing and developing primary center. Bone growth in a shaft begins where bone-forming cells

Introduction

Figure 1.28 Epiphyseal surface (growth surface) in the human and nonhuman. Notice that the surface is a different texture than that of any other kind of bone surface.

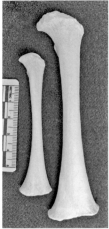

Figure 1.27
Human femur
at two stages of
development.

(osteoblasts) enter the cartilage matrix and begin to secrete a substance that is quickly mineralized. These bone-forming cells enter the matrix through a blood vessel (the nutrient foramen) that leaves a foramen in the completed bone. The position and size of this foramen may help somewhat in identifying a bone.

The area of rapid growth between the diaphysis and epiphysis (or epiphyses) is the growth plate or metaphysis. As the cartilage matrix is turned into bone at the diaphysis and epiphysis, the cartilage between the two continues to grow and add new cells. In this way the bone growth between the diaphysis and epiphysis can continue. When the bone formation at the diaphysis meets the formation of bone at the epiphysis, the two unite and longitudinal bone growth ceases (the bone will not grow longer). This union can occur at different times in different bones, and the sequence and degree of union are useful in the determination of age at death. Also, if this happens too quickly, the bones may be shorter than normal (Figure 1.30). If it happens too late, the bone may be longer than normal. This massive difference in size caused by accelerated or retarded union of growth centers may be a cause for confusion in species identification.

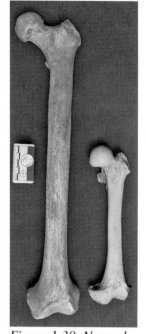
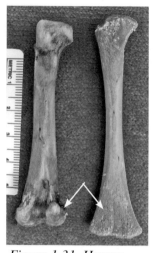

Figure 1.29 Different growth centers in the human femur.

Figure 1.30 Normal femur (left) and shortened femur growth in an achondroplastic dwarf (right). This may be confusing in species identification.

Figure 1.31 Human infant femur (right) and immature chicken femur (left). Though very similar, and both have porous bone, note that the distal articular surfaces are quite different.

Because some of the evidence may be very small, when collecting evidence (including human remains) from a scene it is extremely important to document the location of and to collect every small bit of bone. If this is not done, the process of identification of human remains and the circumstances surrounding the death of that individual may be compromised.

Initial bone formation occurs very quickly and produces loosely woven bone (more a collection or weaving of spicules of bone). A significant

Introduction 29

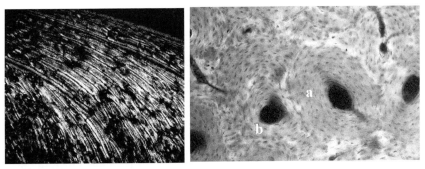

Figure 1.32 Microscopic cross sections of deer bone (left) showing plexiform bone and human bone (right) showing round osteons. Notice that "a" is a complete osteon and "b" is an incomplete osteon.

amount of cartilage remains within these areas of rapid bone growth, and if dry bone (in which much of the organic component is removed) is observed at this stage, it will appear to be very porous (Figure 1.31). This bone growth is so rapid that it traps osteoblasts, which then become osteocytes (bone cells).

Microscopically, these bone cells are important for age determination as well as (often) species designation. As an individual ages, bone is constantly remodeled when bone-absorbing cells (osteoclasts) remove calcium in tunnels they create through the bone cells. New bone is formed in these tunnels but it overlaps the older bone-cell systems (called osteons), creating fragmentary osteons. Relative age can be determined by counting the complete (younger) osteons and the fragmentary (older) osteons. Notice in Figure 1.32 that "a" is a complete, and therefore newer osteon, while "b" is an incomplete, and therefore older osteon.

Many species have bone cell systems that differ from those in human bone. Artiodactyls, for example, have osteons that are more in the shape of curved bricks (or *plexiform* bone) instead of the round pattern seen in humans (see Figure 1.32). Some nonhuman animals have round osteons, but the rest of the bone differs in microscopic and macroscopic morphology. This histological examination is within the realm of expert diagnosis, however, and will not be discussed in detail in this book.

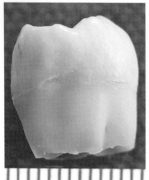

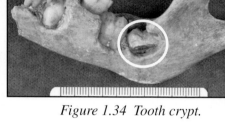

Figure 1.34 Tooth crypt.

Figure 1.33 Developing human molar. Dentition forms from the crown to the root.

Figure 1.35 Cutaway mandible showing developing dentition.

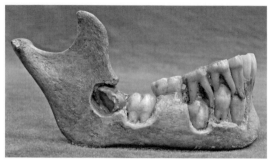

Dental Growth, Development, and Eruption

There are two sets of teeth in mammals: the deciduous (or baby) teeth followed by the permanent dentition. All teeth develop from the crown to the tip (apex) of the root (Figure 1.33), and begin to develop deep inside the maxilla or mandible before reaching a point in their development at which they erupt beyond the gum line (or the alveolus in bone) and become noticeable in the mouth. Note that in a young mandible, there are large voids (called crypts) in which the crown of the tooth develops, and those voids close around the root of the tooth as it erupts beyond the alveolus. If you find that large void and you are certain that it was not caused by a disease process such as an abscess, you know that a tooth was developing in that void (Figure 1.34).

Beneath (in the mandible) or above (in the maxilla) the deciduous tooth, the permanent tooth is forming, and while it is expanding in size, the root of the deciduous tooth begins to resorb. When enough of the deciduous root is gone, the tooth falls out (Figure 1.35).

Introduction

Special Notes about Bird Skeletons

This field guide is primarily a guide to humans and to nonhuman mammals. Human bones are, of course, also confused with bird bones as well as some amphibians, reptiles and even some large fish! Immature humans are the most frequently confused with birds, and other non-mammals because immature humans and many of the other species have articular surfaces that are not as sculpted as mature or even immature mammals.

This book is not intended to be a complete guide for human skeletons compared to birds, reptiles and amphibians, but a few notes about bird skeletons are appropriate.

Almost everyone knows that bird skeletons are very lightweight and have a large marrow cavity (in other words, the cortical, or compact bone is small relative to the cross-section size of the bone. Figure 1.36 shows the cross section of a turkey femur (left) and a large mammal (a cow). Bird bones are actually pneumatic in that they are part of the respiratory system.

Figure 1.36 Turkey femur cross section (left) and cow femur cross section (right).

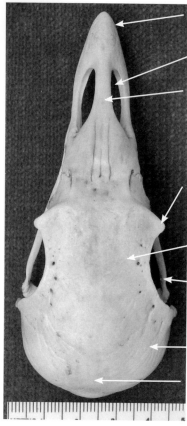

Figure 1.37 Turkey vulture (C. aura septentrionalis) *cranium superior view.*

Figure 1.38 Turkey vulture (C. aura septentrionalis) *cranium lateral view.*

Introduction

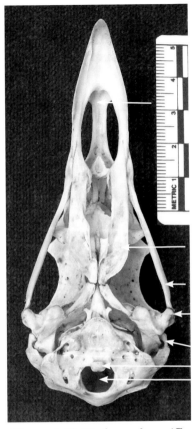

Figure 1.39 Turkey vulture (C. aura septentrionalis) cranium inferior view.

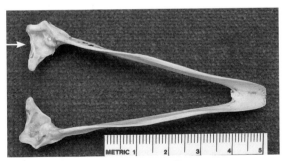

Figure 1.40 Turkey vulture (C. aura septentrionalis) dentary (mandible) superior view.

Features of a Bird Skull

The following photographs represent skulls of a few North American raptors. They illustrate some features common to most bird skulls, although birds of prey do show skeletal features that differ from other bird types.

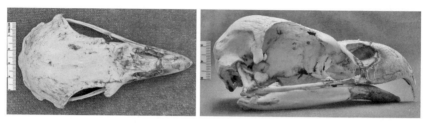

Figure 1.41 Bald eagle (Haliaeetus leucocephalus).

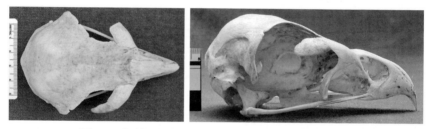

Figure 1.42 Golden eagle (Aquila chrysaetos).

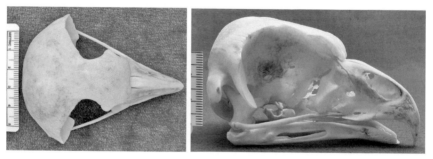

Figure 1.43 Great horned owl (Bubo virginianus).

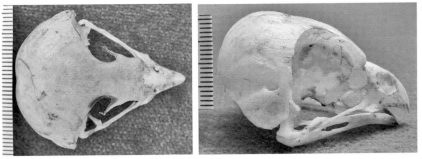

Figure 1.44 Burrowing owl (Athene cunicularia).

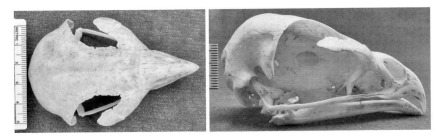

Figure 1.45 Red-tailed hawk (Buteo jamaicensis).

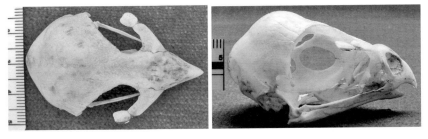

Figure 1.46 Cooper's hawk (Accipiter cooperii)

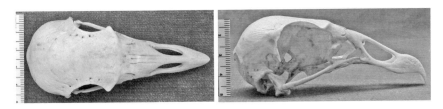

Figure 1.47 Turkey vulture (Cathartes aura septentrionalis).

Postcranial Bird Skeletal Elements

These are photographs of the postcranial elements of the domestic turkey (*Meleagris* sp.). The general characteristics are common to many birds, but, depending upon the lifestyles of various birds, the specific morphological characteristics vary considerably. See Figure 1.4 for a labeled skeleton of a chicken.

Figure 1.49 Scapula blade. Arrow points to glenoid fossa for humerus.

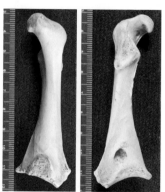

Figure 1.50 Coracoid bone (analogous to our coracoid process) of scapula.

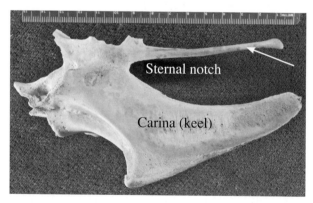

Figure 1.51 Sternum (anterior aspect at left). Most birds have a sternum with a keel as shown above, but some flightless birds have a reduced sternum. Arrow points to the lateral caudal process (or external lateral xiphoid).

Introduction

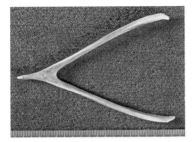

Figure 1.52 Turkey furcula (fused clavicles) commonly called the wishbone.

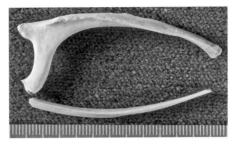

Figure 1.53 Vertebral rib (top) and sternal rib.

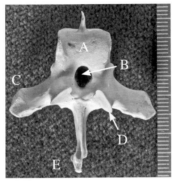

A: Centrum (body)
B: Neural canal
C: Transverse process
D: Articular facet
E: Spinous process

Figure 1.54 Turkey thoracic vertebra.

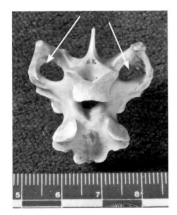

Figure 1.55 Turkey cervical vertebra. Note foramina (arrows).

Figure 1.56 Turkey fused thoracic vertebrae.

Figure 1.57 Synsacrum comprised of the three lumbar, seven sacral, and the first six caudal vertebrae. The remaining six caudal vertebrae form the tail (not shown), ending with a pygostyle which is a section of fused caudal vertebrae). Upper photograph is dorsal view, lower photograph is lateral view.

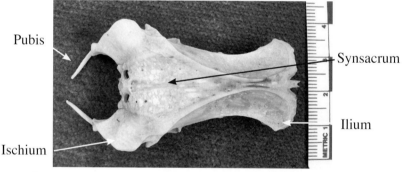

Figure 1.58 Red-shouldered hawk (Buteo lineatus). Dorsal view of pelvic girdle with synsacrum and ox coxae.

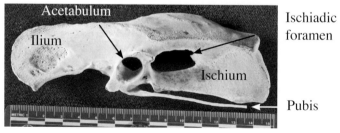

Figure 1.59 Turkey left os coxa.

Introduction

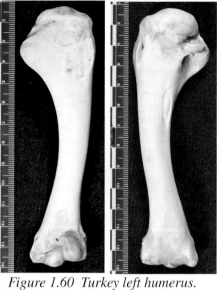

Figure 1.60 Turkey left humerus.

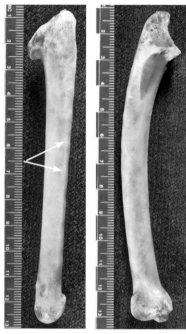

Figure 1.61 Turkey left ulna. Arrows point to two of the quill knobs.

Figure 1.63 Turkey radius head.

Figure 1.62 Turkey radius.

Figure 1.64 Red-shouldered hawk (Buteo lineatus) bones at end of wing.

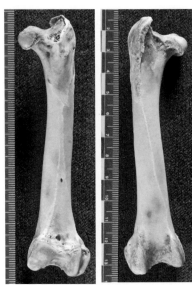

Figure 1.65 Turkey femur.

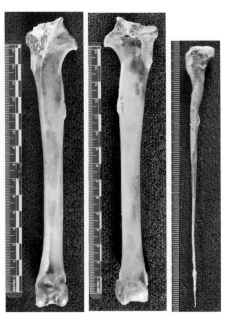

Figure 1.66 Turkey tibiotarsus (left and center) and fibula (right).

Figure 1.67 Tarsometatarsus of red-shouldered hawk (Buteo lineatus) tarsometarsus (left) and immature chicken (right).

Cleaning and Storing Skeletal Elements

Be sure you understand the laws regarding collection of wildlife remains in your area! The U.S. Fish and Wildlife Service imposes stiff fines for collecting the remains of many birds and other wild animals *even if that animal is already dead before you arrive on the scene.*

The information provided here is not intended to encourage the unlawful collection of wildlife remains for any purpose!

The following suggestions are for cleaning and storing nonhuman skeletal material. It should be obvious that collection, cleaning and storing human remains should only be done by appropriately trained authorities under law enforcement and medical examiner standards. If private citizens discover human remains (or remains not *known* to be nonhuman), nothing should be touched, and the area should be vacated using the same way out as was used to get into the scene (so that evidence is not further disturbed). Proper law enforcement authorities must be contacted *immediately*.

The goal in cleaning skeletal elements is, of course, to completely remove the soft tissue off of the skeleton without damaging the bone. Remaining soft tissue will usually continue to decompose (depending upon the moisture content in the soft tissue and the environment in which they are stored), although desiccated soft tissue stored in a very dry environment can remain relatively unaltered for years.

In general, the more slowly soft tissue is removed, the less likely the bone will be damaged. Intense heat and harsh chemicals can destroy the organic and inorganic components of bone and will make the bone weaker (and even destroy some of the features).

Dermestid Beetles

One of the slowest and most gentle ways to clean skeletal elements is by dermestid beetles. These small beetles will effectively remove the soft tissue but leave the bone clean and intact. However, most species of dermestid beetles feed on dry soft tissue, so if the soft tissue is still in active wet decomposition, the beetles usually will not eat it. In addition, if the soft tissue is wet, maggots, bacteria, and other organisms can compete with the dermestids or actually infect the dermestids with disease. In addition, dermestid beetles must be well contained, as they will destroy other materials such as natural fibers in homes and businesses.

Water Cleaning Methods

Other than dermestid beetles, water-based methods are the safest, gentlest way to clean soft tissue from skeletal elements. Place the skeletal element(s) into a solution of a half cup of laundry detergent or dishwashing soap to about a gallon of water. Completely cover the elements with the solution and gently heat to about 100 to 150 degrees Fahrenheit. Check every few hours (or overnight), remove loose soft tissue and return to a fresh solution until the soft tissue is completely gone (this may take several days, but it is important to check the bones frequently).

This method removes the soft tissue from the outside of the remains, but it does not remove the marrow and other soft tissue from inside the bones. In specimens that are intended for permanent curation in museums, small holes are often drilled in the ends of long bones, and the bones are placed back in the water solution to remove the fats and other soft tissue from the inside. Often, as a finishing touch, a small amount of bleach may be added to a final clean water bath.

Drying and Storing Skeletal Elements

After removing the skeletal elements from the water bath, spread the bones on a clean sheet with an absorbent material (such as paper towels, additional sheets, clean diapers, etc.) underneath. Allow the elements to dry *slowly* out of direct heat or sunlight. If bones are allowed to dry too quickly, the outer surface dries more quickly than does the inner surface, and the outer surface will crack. As the bone dries, these cracks will become larger. If the bones are allowed to dry slowly, cracks will rarely occur.

Cautionary Notes:

- Never put wet bones in a plastic bag or other airtight container; mold will grow and contribute to the destruction of the bones themselves.

- Never put skeletal elements that still have soft tissue attached in a refrigerator for longer than about a week, as the soft tissue will continue to decompose (just like other meat). If you cannot clean the remains immediately, freeze them until you are ready.

- Store dry skeletal elements in a dry location.

Preserving Skeletal Elements

If skeletal elements are dry and stored in a dry location, they rarely need an application of polyurethane, varnish, or other chemical. If polyurethane or varnish is to be applied, be sure that the bone is completely dry and free of oils, or the applied material will not adhere to the bone.

Part II

Major Bones of the Bodies of Different Animals

Cranium

Features of the cranium 49
    Human (*Homo sapiens*) 50
    Moose (*Alces alces*) 51
    Elk (*Cervus elaphus*) 51
    Deer (*Odocoileus* sp.) 52
    Bison (*Bison bison*) 52
    Cow (*Bos taurus*) 53
    Antelope (*Antilocapra americana*) 53
    Mountain sheep (*Ovis canadensis*) 54
    Domestic sheep (*Ovis aries*) 54
    Domestic pig (*Sus scrofa*) 55
    Llama (*Lama glama*) 55
    Horse (*Equus*) 56
    Bear (*Ursus americanus*) 56
    Wolf (*Canis lupus*) 57
    Coyote (*Canis latrans*) 57
    Domestic dog (*Canis domesticus*) 57
    Mountain lion (*Felis concolor*) 58
    Bobcat (*Lynx rufus*) 58
    Domestic cat (*Felis domesticus*) 58
    Raccoon (*Procyon lotor*) 59
    Badger (*Taxidea taxus*) 59
    Skunk (*Mephitis mephitis*) 60
    River otter (*Lontra canadensis*) 60
    Rabbit (*Lepus* sp.) 61
    Beaver (*Castor canadensis*) 61
    Porcupine (*Erethizon dorsatum*) 62
    Marmot (*Marmota monax*) 62
    Prairie dog (*Cynomys gunnisoni*) 63
    Norway rat (*Rattus norvegicus*) 63
    Squirrel (*Sciuridae sciurus niger*)........................... 64
    Armadillo (*Dasypus novemcinctus*) 64
    Opossum (*Didelphis virginiana*) 65
    Seal (*Phoca vitulina*) 65

Features of the Cranium

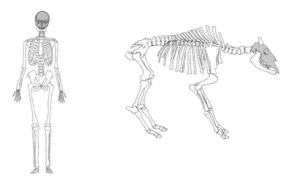

The skull is the cranium plus the mandible.

The cranium articulates with the first cervical vertebra by means of the occipital condyles, which lie on the periphery of the foramen magnum. The spinal cord passes through the foramen magnum on its way from the brain to the vertebral column. Note the placement of the foramen magnum in humans and nonhumans. In nonhumans the foramen magnum is usually placed more posteriorly.

The posterior neck muscles insert on the cranium behind the foramen magnum. In humans the uppermost point of insertion is a raised line on the posterior portion of the occipital bone called the superior nuchal line or sometimes the superior nuchal crest (though the latter term is more often used with nonhumans, as many times that line is so raised to form a crest).

One large anterior neck muscle in humans inserts on the mastoid process, which is larger in humans than in most other animals.

Although there is variation, the general shape of the cranium (even in fragments) along with the placement of the sutures and foramina can help determine whether the fragment is from a human or nonhuman.

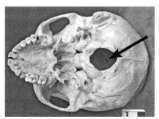 

Note the posterior placement of foramen magnum in nonhuman.

Human Cranial Features

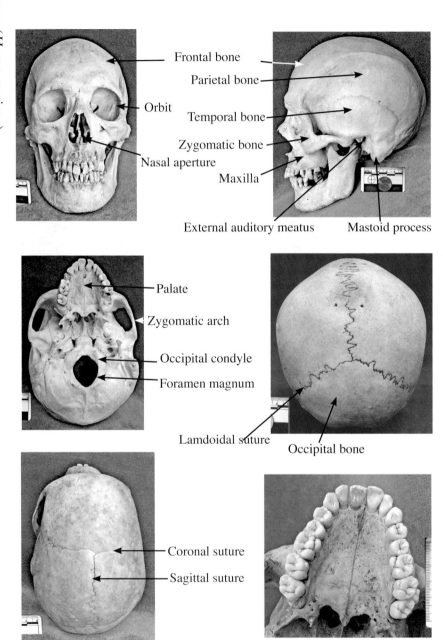

Cranium

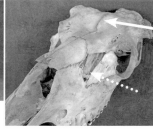

Moose (*Alces alces*)

Note no upper incisors, fenestration (window on maxilla - dotted arrow) and "bump" between antlers (black arrow).

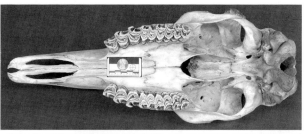

Note the long nasal extension.

As with the moose, expect variation due to the sex of the animal.

Lacrimal pits (arrow above)

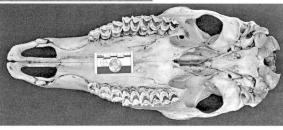

Elk (*Cervus elaphus*)

Deer (*Odocoileus* sp.)

Notice fenestra (window) at arrow. Also notice pillared cheek teeth and no anterior teeth on maxilla.

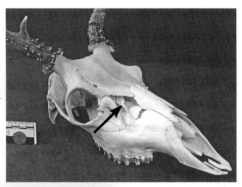

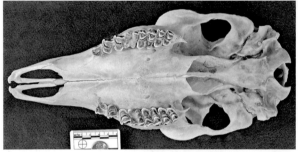

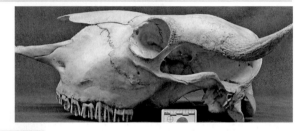

No fenestra, no anterior maxillary teeth. Buttressed, pillared cheek teeth.

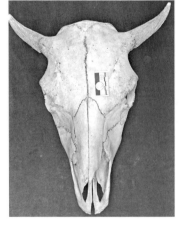

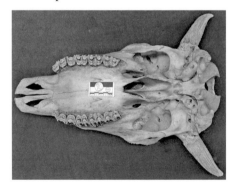

Bison (*Bison bison*)

Cranium

No fenestra (window), no anterior maxillary teeth.

Notice buttressed, pillared cheek teeth.

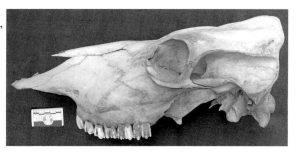
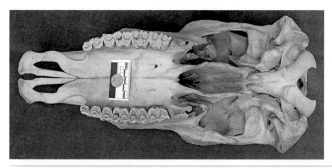

Cow (*Bos taurus*)

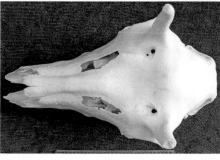
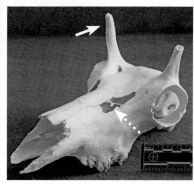

Knife-like horn core (white arrow).

Large, extended round orbit, fenestra (dotted arrow), and foramen at base of horn core.

Antelope (*Antilocapra americana*)

Mountain sheep (*Ovis canadensis*)

No fenestra (window anterior to orbit).

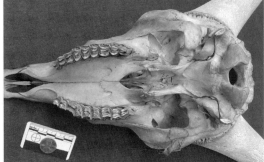

No anterior teeth on maxilla, buttressed, pillared cheek teeth.

Domestic sheep (*Ovis aries*)

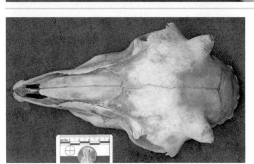

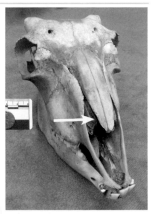

Long nasals (arrow).

No fenestra, no anterior maxillary teeth.

Cranium

Note angles at posterior cranium

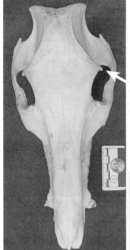

No fenestra

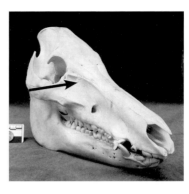

Postorbital projections

Crenulated molars

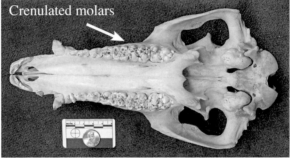

Domestic pig (*Sus scrofa*)

Diamond-shaped cranium

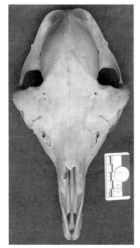

Small fenestra

No incisors on maxilla

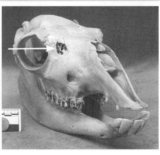

Long nasals

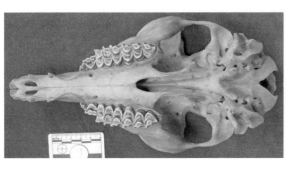

Llama (*Lama glama*)

Horse (*Equus*)

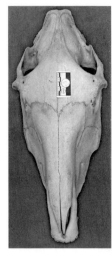

Long nasal extensions

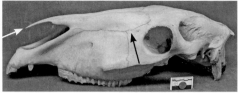

No fenestra

Anterior maxillary dentition present

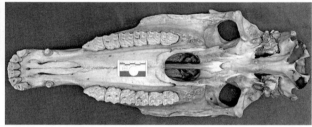

Bear (*Ursus americanus*)

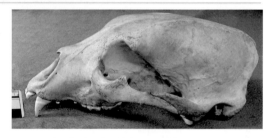

Relatively short nasals and gradual slope on frontal

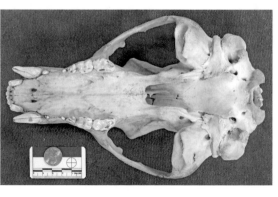

Elongated crenulated molars

Cranium

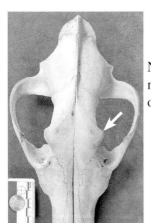

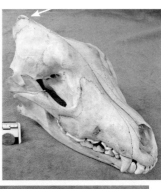

Sagittal crest

No fenestra, long narrow snout, post-orbital projections

Shearing teeth, long canines

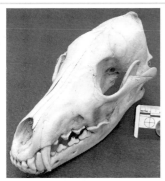

Coyote (*Canis latrans*)

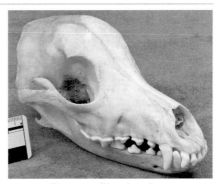

German Shepherd

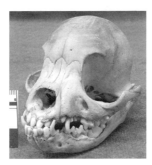

Pug

Wolf (*Canis lupus*)

Coyote (*Canis latrans*), Domestic dog (*Canis domesticus*)

Mountain lion (*Felis concolor*)

Rounded, short cranium, large canines

Postorbital projections

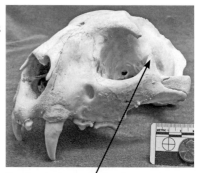

Projection on malar bone to more complete orbit than in dogs

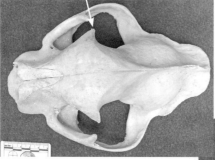

Triangular, short palate with only two cheek teeth

Curled, locking temperomandibular joint

Bobcat (*Lynx rufus*), Domestic cat

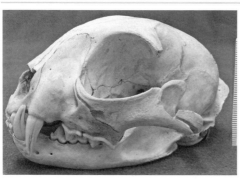

Bobcat (*Lynx rufus*)

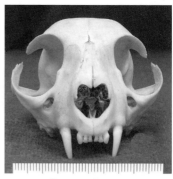

Domestic cat (*Felis domesticus*)

Cranium

Curve on frontal bone and nasals

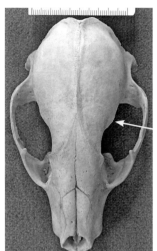

No postorbital projections

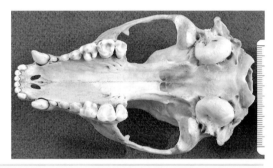

Raccoon (*Procyon lotor*)

Note shape of posterior cranium

Angled frontal area

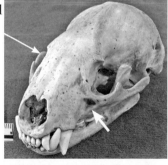

Triangular infraorbital foramen

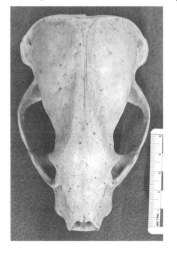

Large bulla (contains middle and inner ear)

Badger (*Taxidea taxus*)

Skunk (*Mephitis mephitis*)

Short face, reduced cheek teeth

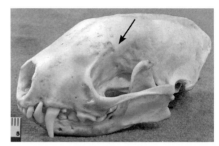

No postorbital projections

River otter (*Lontra canadensis*)

Long infraorbital foramen

Projections posterior to external auditory meatus

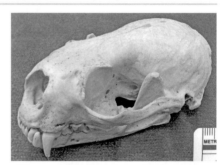

Short nasal extensions, flat cranium.

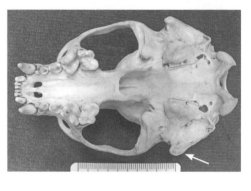

Cranium

Two sets of maxillary incisors

Fenestra (window) on maxilla

Long postorbital projections

"H"-shaped cheek teeth

Rabbit (*Lepus* sp.)

Long projections behind ear

Long incisors that continue to grow

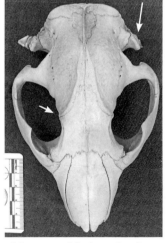

No postorbital projection

Deep notch

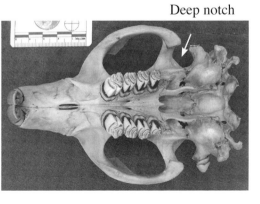

Beaver (*Castor canadensis*)

Porcupine (*Erethizon dorsatum*)

No postorbital projection

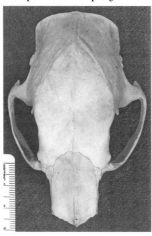

Large nasal opening

Large infraorbital foramen

Large incisors continue to grow

Rounded sagittal crest

Note morphology above incisors

Flat cranium

Marmot (*Marmota monax*)

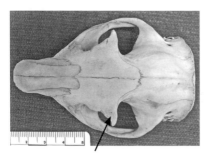

Postorbital projections

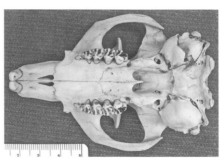

Cranium

Postorbital projections

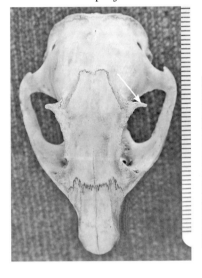

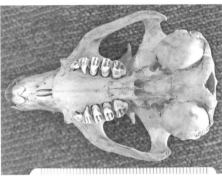

Tooth row converges toward back

Prairie dog (*Cynomys gunnisoni*)

Large infraorbital foramen

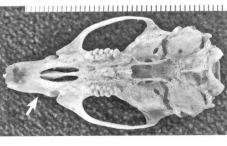

Large diastema (gap)

Norway rat (*Rattus norvegicus*)

Squirrel (*Sciuridae sciurus niger*)

Postorbital projections

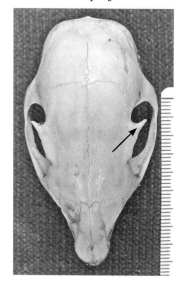

Note angle at nasal opening

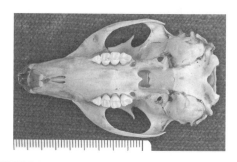

Armadillo (*Dasypus novemcinctus*)

No anterior teeth on mandible or maxilla

Long, narrow nasal region

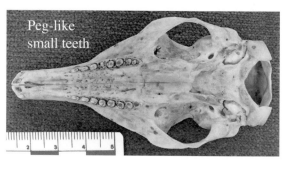

Peg-like small teeth

Cranium

Small postorbital projections

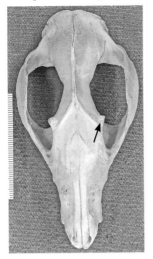

Note sagittal crest

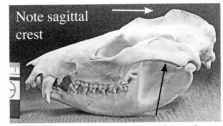

Wide zygomatic arch

Note tooth form (particularly the incisors)

Opossum (*Didelphis virginiana*)

Straight slope to forehead

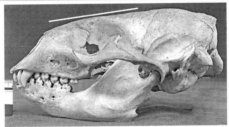

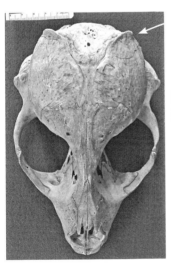

Rounded, bulbous cranium from superior view

Small pointed projections

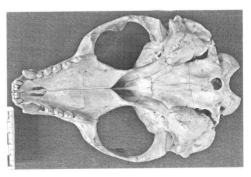

Seal (*Phoca vitulina*)

Mandible

Features of the mandible 69
    Human (*Homo sapiens*) 70
    Moose (*Alces alces*) 71
    Elk (*Cervus elaphus*) 71
    Deer (*Odocoileus* sp.) 72
    Bison (*Bison bison*) 72
    Cow (*Bos taurus*) 73
    Antelope (*Antilocapra americana*) 73
    Mountain sheep (*Ovis canadensis*) 74
    Domestic sheep (*Ovis aries*) 74
    Domestic pig (*Sus scrofa*) 75
    Llama (*Lama glama*) 75
    Horse (*Equus*) 76
    Bear (*Ursus americanus*) 76
    Wolf (*Canis lupus*) 77
    Coyote (*Canis latrans*) 77
    Mountain lion (*Felis concolor*) 78
    Bobcat (*Lynx rufus*) 78
    Raccoon (*Procyon lotor*) 79
    Badger (*Taxidea taxus*) 79
    Skunk (*Mephitis mephitis*) 80
    River otter (*Lontra canadensis*) 80
    Rabbit (*Lepus* sp.) 81
    Beaver (*Castor canadensis*) 81
    Porcupine (*Erethizon dorsatum*) 82
    Marmot (*Marmota monax*) 82
    Prairie dog (*Cynomys gunnisoni*) 83
    Norway rat (*Rattus norvegicus*) 83
    Squirrel (*Sciuridae sciurus niger*) 84
    Armadillo (*Dasypus novemcinctus*) 84
    Opossum (*Didelphis virginiana*) 85
    Seal (*Phoca vitulina*) 85

Features of the Mandible

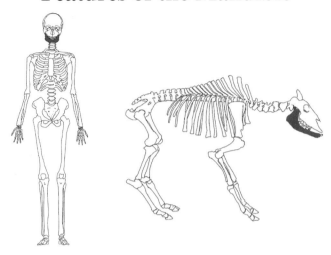

The shape and size of the mandible are influenced by, among other things the size and form of the dentition, which, of course, reflects the diet of the animal, and the size of the nasal region, which reflects the importance of the sense of smell. Most animals that survive on vegetation have flatter cheek teeth for grinding while carnivores have dentition that grabs and tears tissue.

The mandible (lower jaw) articulates with the cranium at the temporomandibular joint by its mandibular condyles. The condyles and the coronoid processes are at the superior aspect of the ascending ramus (the coronoid process is the area of insertion of the masseter muscle, one of the muscles used in chewing). The angle, breadth and height of the ascending ramus are different in different animals. Unlike other animals, the human mandible has a chin.

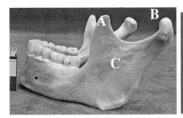
Human (*Homo sapiens*)

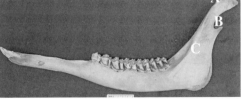
Moose (*Alces alces*)

A: Coronoid Process B: Mandibular Condyles C: Ascending Ramus

Human (*Homo sapiens*)

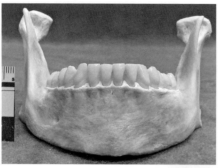

The dental arcade in humans is generally shorter than in most quadrupeds, though there is some variation in shape.

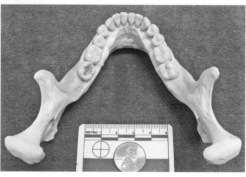

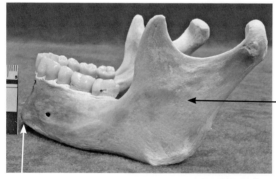 ← Wide ascending ramus

Chin

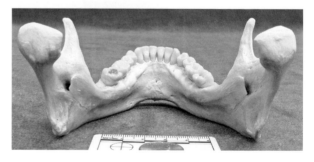

Mandible

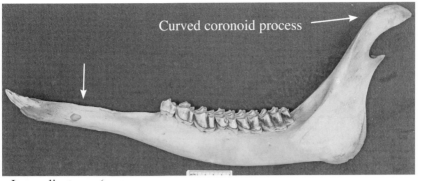

Moose (*Alces alces*)

Curved coronoid process

Long diastema (gap between cheek teeth and incisors)

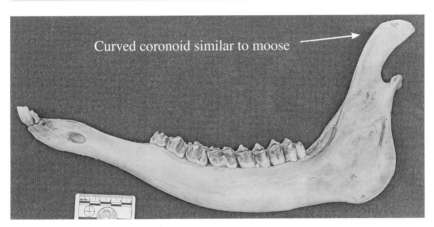

Curved coronoid similar to moose

Elk (*Cervus elaphus*)

Deer (*Odocoileus* sp.)

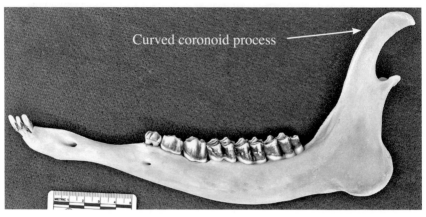

Curved coronoid process

Overall, the deer mandible is smaller and more delicate than those of moose and elk.

Bison (*Bison bison*)

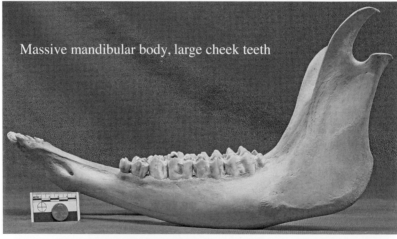

Massive mandibular body, large cheek teeth

Mandible

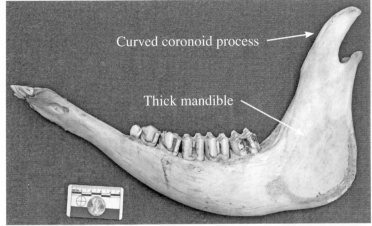

Curved coronoid process

Thick mandible

Cow (*Bos taurus*)

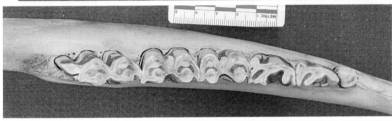

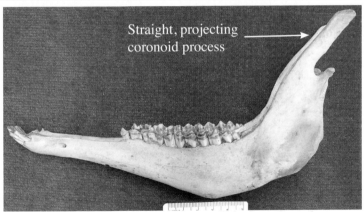

Straight, projecting coronoid process

The antelope has a narrow mandible from lateral edge to lateral edge (viewed from above).

Antelope (*Antilocapra americana*)

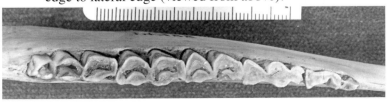

74　　　　　　　　　　　　　　　　　　　Human and Nonhuman Bone Identification

Mountain sheep (*Ovis canadensis*)

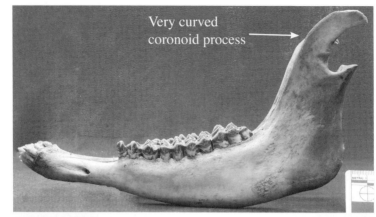

Very curved coronoid process

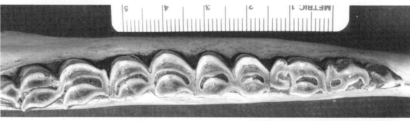

Domestic sheep (*Ovis aries*)

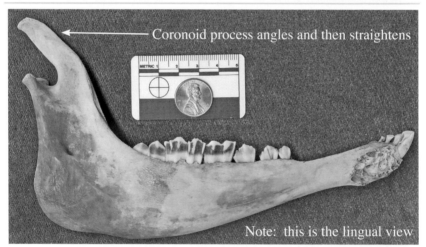

Coronoid process angles and then straightens

Note: this is the lingual view

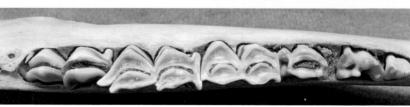

Mandible

Blocky ascending ramus with short coronoid process

Crenulated cheek teeth

Domestic pig (*Sus scrofa*)

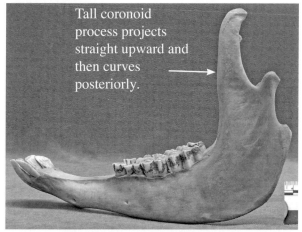

Tall coronoid process projects straight upward and then curves posteriorly.

Coronoid process curves medially

Llama (*Lama glama*)

76 Human and Nonhuman Bone Identification

Horse (Equus)

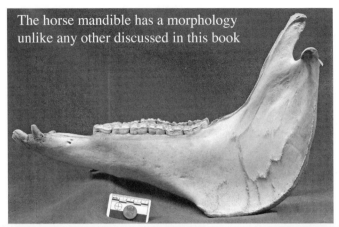

The horse mandible has a morphology unlike any other discussed in this book

Sometimes the canines are missing

Bear (Ursus americanus)

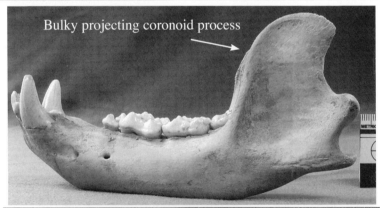

Bulky projecting coronoid process

Molars are crenulated until worn, and then are smoother

Mandible

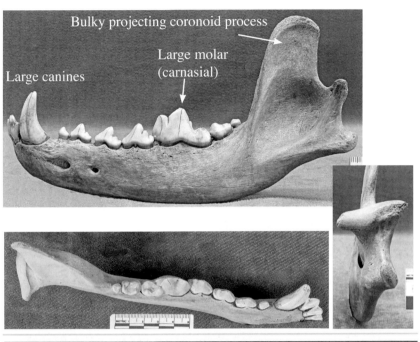

Wolf (*Canis lupus*)

Bulky projecting coronoid process
Large molar (carnasial)
Large canines

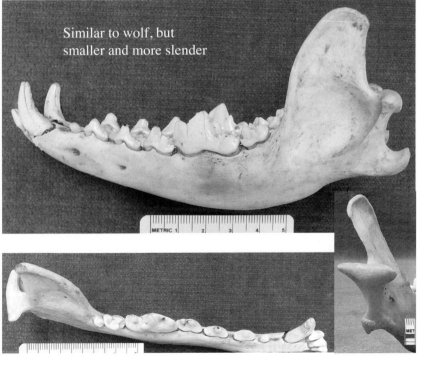

Coyote (*Canis latrans*)

Similar to wolf, but smaller and more slender

78　Human and Nonhuman Bone Identification

Mountain lion (*Felis concolor*)

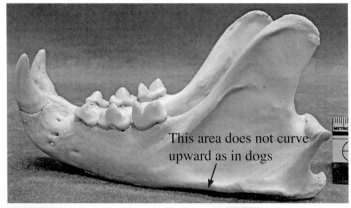

This area does not curve upward as in dogs

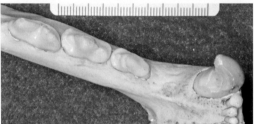

Cats have shorter mandibles than do dogs, and have fewer teeth.

Bobcat (*Lynx rufus*)

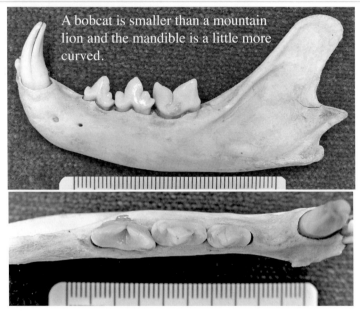

A bobcat is smaller than a mountain lion and the mandible is a little more curved.

The reduction in the length of the mandible and the reduction in the number of teeth is characteristic of all cats (including domestic cats).

Mandible

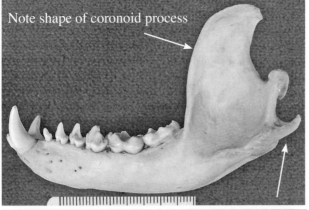

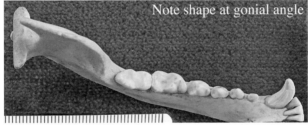

Raccoon (*Procyon lotor*)

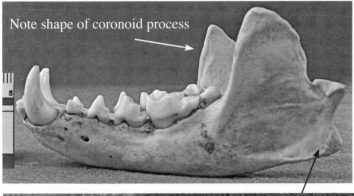

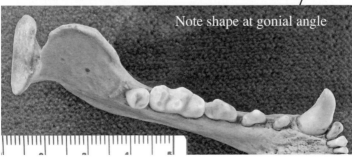

Badger (*Taxidea taxus*)

Skunk (*Mephitis mephitis*)

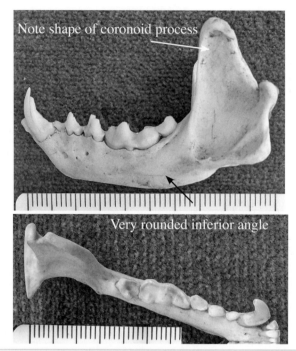

River otter (*Lontra canadensis*)

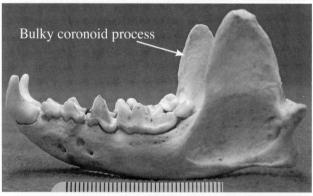

Mandible

Rabbit (*Lepus* sp.)

Note shape at gonial angle that quickly narrows toward anterior aspect

Note shape of cheek teeth

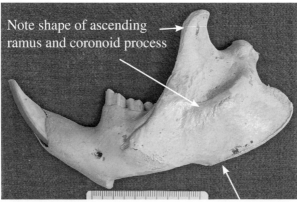

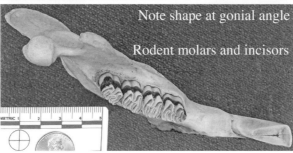

Beaver (*Castor canadensis*)

Note shape of ascending ramus and coronoid process

Note shape at gonial angle

Rodent molars and incisors

Porcupine (*Erethizon dorsatum*)

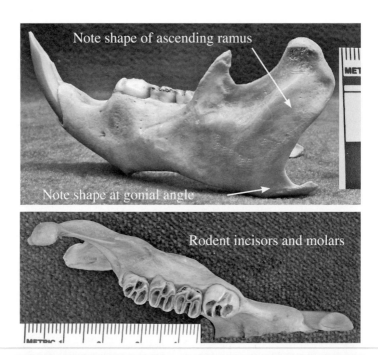

Marmot (*Marmota monax*)

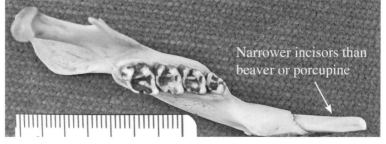

Mandible

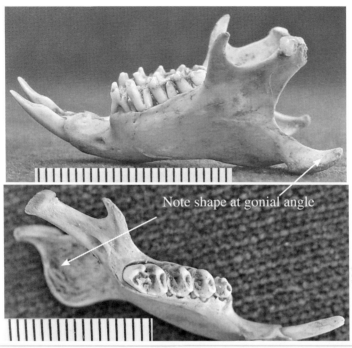

Prairie dog (*Cynomys gunnisoni*)

Note shape at gonial angle

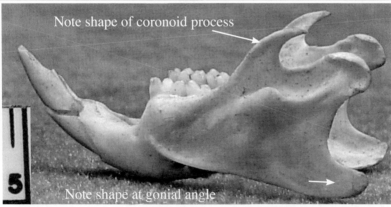

Note shape of coronoid process

Note shape at gonial angle

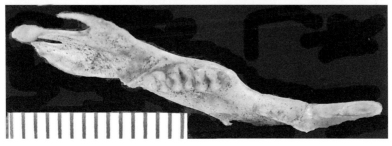

Norway rat (*Rattus norvegicus*)

Squirrel (*Sciuridae sciurus niger*)

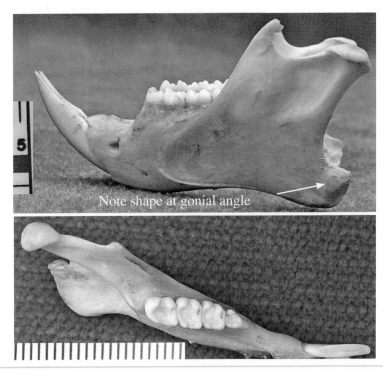

Note shape at gonial angle

Armadillo (*Dasypus novemcinctus*)

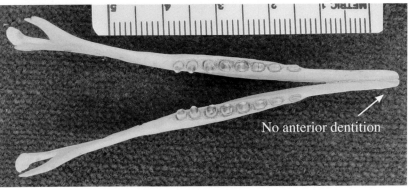

No anterior dentition

Long, narrow, thin-boned mandible with peg-like teeth

Mandible

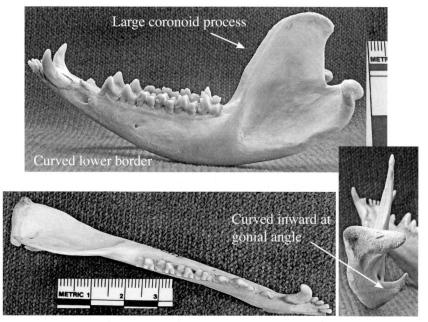

Opossum (*Didelphis virginiana*)

- Large coronoid process
- Curved lower border
- Curved inward at gonial angle

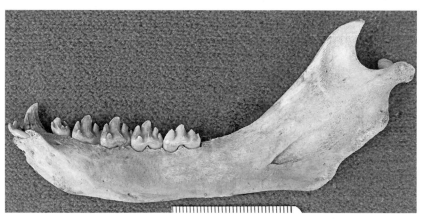

Note: this is the lingual view

Seal (*Phoca vitulina*)

Scapula

Features of the scapula 89
    Human (*Homo sapiens*) 90
    Moose (*Alces alces*) 91
    Elk (*Cervus elaphus*) 91
    Deer (*Odocoileus* sp.) 92
    Bison (*Bison bison*) 92
    Cow (*Bos taurus*) 93
    Antelope (*Antilocapra americana*) 93
    Mountain sheep (*Ovis canadensis*) 94
    Domestic sheep (*Ovis aries*) 94
    Domestic pig (*Sus scrofa*) 95
    Llama (*Lama glama*) 95
    Horse (*Equus*) 96
    Bear (*Ursus americanus*) 96
    Wolf (*Canis lupus*) 97
    Coyote (*Canis latrans*) 97
    Mountain lion (*Felis concolor*) 98
    Bobcat (*Lynx rufus*) 98
    Raccoon (*Procyon lotor*) 99
    Badger (*Taxidea taxus*) 99
    Skunk (*Mephitis mephitis*) 100
    River otter (*Lontra canadensis*) 100
    Rabbit (*Lepus* sp.) 101
    Beaver (*Castor canadensis*) 101
    Porcupine (*Erethizon dorsatum*) 102
    Marmot (*Marmota monax*) 102
    Prairie dog (*Cynomys gunnisoni*) 103
    Norway rat (*Rattus norvegicus*) 103
    Squirrel (*Sciuridae sciurus niger*) 104
    Armadillo (*Dasypus novemcinctus*) 104
    Opossum (*Didelphis virginiana*) 105
    Seal (*Phoca vitulina*) 105

Features of the Scapula

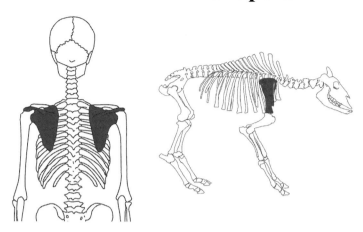

The scapula is commonly called the "shoulder blade" and rests on the upper back in humans and on the dorsolateral surface in the nonhumans in this book. The scapular spine is an area of muscle insertion and generally "points" to the glenoid fossa, which articulates with the humerus. The scapula is triangular in most mammals, but in humans the spine is diagonal to the bone when it is in anatomical position (though it still "points" to the glenoid fossa). In many nonhuman mammals the scapula is elongated.

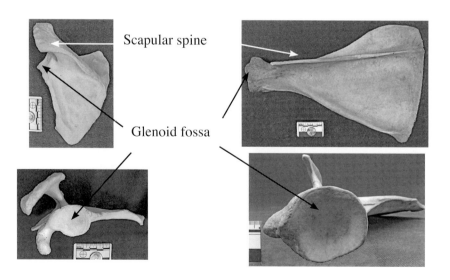

Human (*Homo sapiens*)

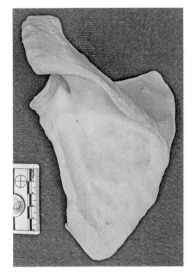

Note triangular shape

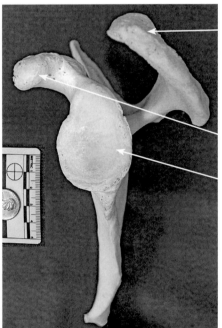

Very strong acromion process

Curved, knobby coracoid process

Acromion process

Coracoid process

Glenoid fossa

Scapula

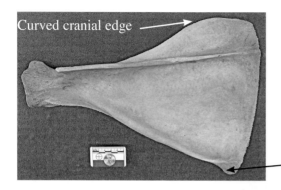

Curved cranial edge

Knobby projection

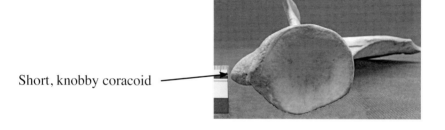

Short, knobby coracoid

Moose (*Alces alces*)

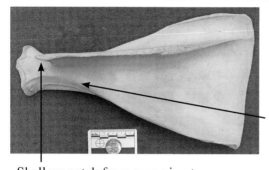

Long scapular neck

Shallow notch from acromion to base of spine

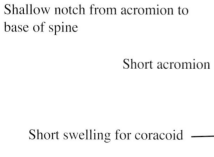

Short acromion

Short swelling for coracoid

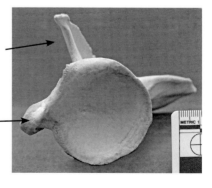

Elk (*Cervus elaphus*)

Deer (*Odocoileus* sp.)

Cranial edge more rounded in deer and more angled in sheep and goats.

Acromion process extends to just short of glenoid fossa.

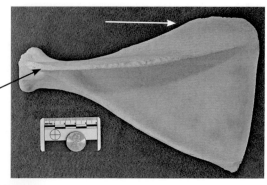

Coracoid process short and curved

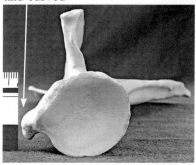

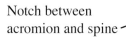

Notch between acromion and spine

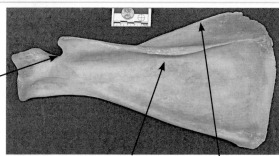

Spine located near cranial edge and is curled near midline

Fairly straight cranial edge

Bison (*Bison bison*)

Short coracoid process

Scapula

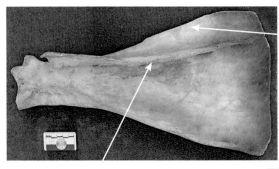

Fairly straight cranial edge

Spine located near cranial edge

Very short coracoid process

Cow (*Bos taurus*)

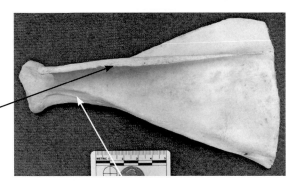

Spine near cranial border

Long neck

Curled coracoid process

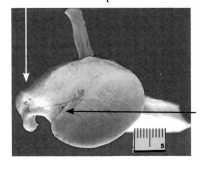

Notch in glenoid fossa

Antelope (*Antiloca pra americana*)

Mountain sheep (*Ovis canadensis*)

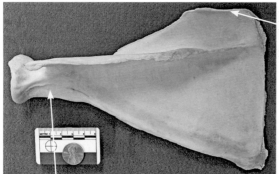

Flat "table" on cranial border with sharp angle medial and lateral

Relatively long neck

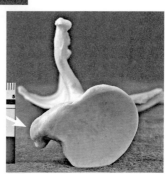

Short, curved coracoid process

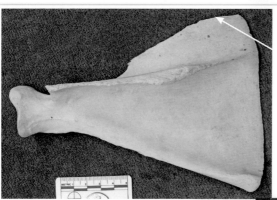

Angle with shorter "table" than in mountain sheep

Domestic sheep (*Ovis aries*)

Triangle shape with long neck

Short, curved coracoid process

Scapula

Note angle on cranial surface

No acromion process

Spine divides scapula into nearly equal upper and lower halves, and caudally turned spine at center of scapula.

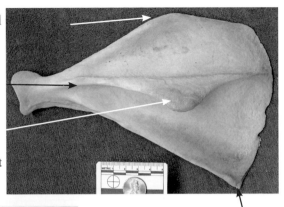

Sharp inferior angle

Domestic pig (Sus scrofa)

Coracoid process small swelling

Angled superior surface

Long concavity medial to glenoid fossa

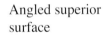
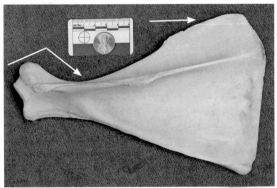

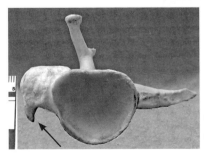

Curved, somewhat pointed coracoid process

Llama (Lama glama)

Horse (*Equus*)

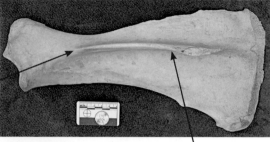

Scapula spine ends well before glenoid fossa.

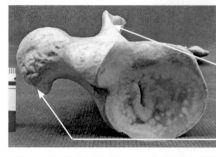

Spine divides scapula almost equally

Very long narrow scapula

No significant acromion process

Pronounced knobby coracoid process

Bear (*Ursus americanus*)

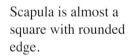

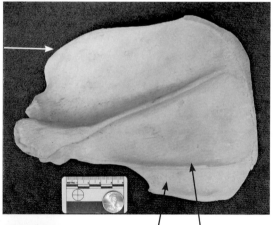

Raised, rounded "hump"

Scapula is almost a square with rounded edge.

Second spine

Fossa created by second spine

Large acromion process

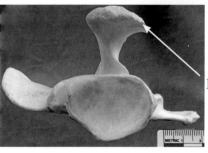

Scapula

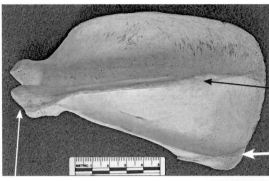

Spine is diagonal.

Buttressed angle

Wolf (Canis lupus)

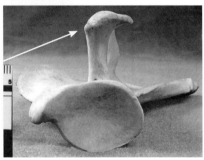

Heavy acromion process (more pronounced than in cats).

Acromion process

Outline is rectangular with rounded edges.

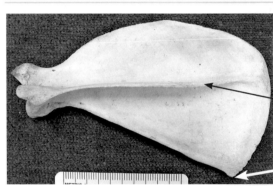

Spine is diagonal.

Buttressed angle

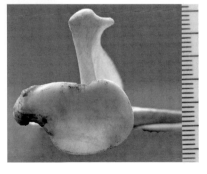

Somewhat more rounded on cranial surface and more triangular on caudal surface than in wolves.

Coyote (Canis latrans)

Mountain lion (*Felis concolor*)

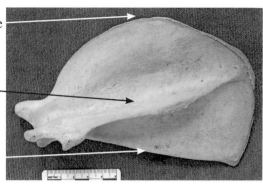

Rounded cranial edge

Spine diagonal with crest turned caudally

Straight caudal edge

Short metacromial process (secondary acromion)

Same morphological features as mountain lion, but smaller

Significant metacromial process (secondary acromion)

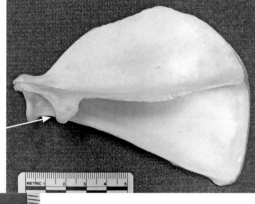

Bobcat (*Lynx rufus*)

Scapula

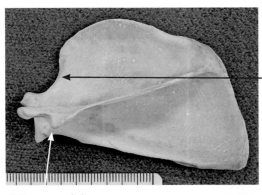

Rounded edge with significant notch medial to glenoid

Metacromial process (second acromion)

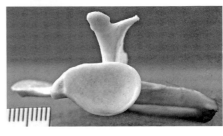

Raccoon (*Procyon lotor*)

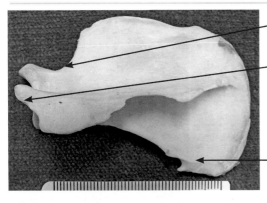

Wide notch

Long acromion process with significant caudal projection

Second fossa

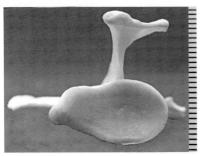

Note rectangular outline

Badger (*Taxidea taxus*)

Skunk (*Mephitis mephitis*)

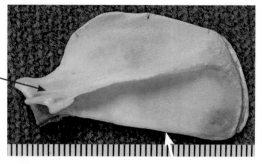

Well-developed acromion process with metacromial process (second acromion)

Straight caudal edge

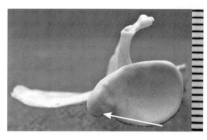

Small bulbous coracoid

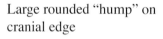

River otter (*Lontra canadensis*)

Large rounded "hump" on cranial edge

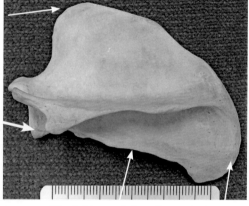

Strong acromion process

Straight caudal edge

Rounded vertebral border

Scapula

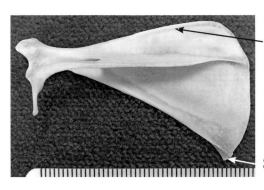

Straight cranial edge

Sharp caudal point

Rabbit (*Lepus* sp.)

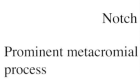

Notch

Prominent metacromial process

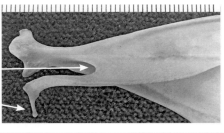

Paddle-shaped blade

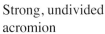

Strong, undivided acromion

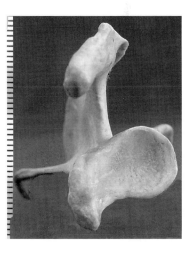

Beaver (*Castor canadensis*)

Porcupine (*Erethizon dorsatum*)

Triangular in shape, rounded cranial edge

Metacromial process (secondary acromion)

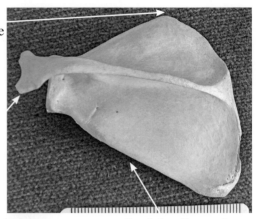

Straight, diagonal caudal edge

Small, blunt coracoid

Marmot (*Marmota monax*)

Wide acromial process with metacromial process secondary acromion)

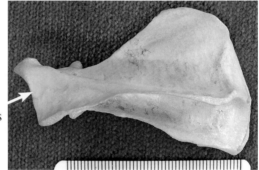

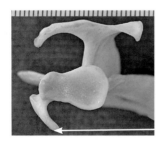

Large coracoid process

Scapula

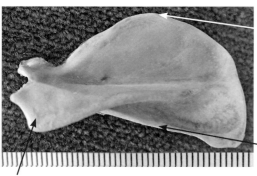

Curved cranial edge

Slightly concave caudal edge

Side acromion process

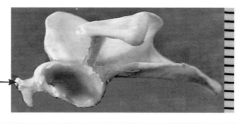

Small, curved coracoid process

Prairie dog (Cynomys gunnisoni)

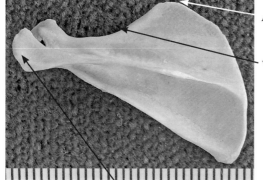

Angled cranial edge

Wide concavity

Strong acromion process

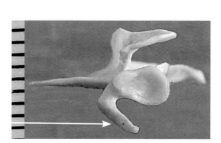

Long, narrow coracoid process

Norway rat (Rattus norvegicus)

Squirrel (*Sciuridae sciurus*)

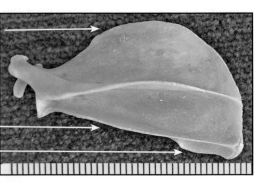

Rounded cranial edge

Straight caudal edge
Squared caudal angle

"D"-shaped scapula

Deep notch between acromion process and spine

Long, curved coracoid process

Armadillo (*Dasypus novemcinctus*)

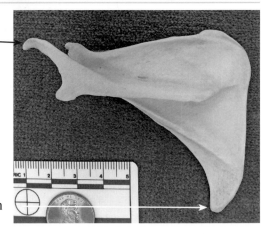

Very long acromion process

Long caudal projection

Short, curved coracoid process

Very long acromion process

Scapula

Oval scapula

Triangular acromion process

Relatively short coracoid process

Opossum (Didelphis virginiana)

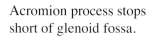

Acromion process stops short of glenoid fossa.

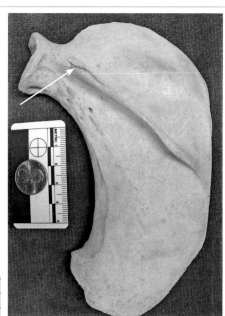

Very short, knobby coracoid process

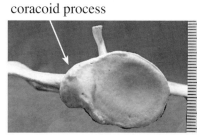

Comma-shaped scapula

Seal (Phoca vitulina)

Humerus

Features of the humerus 109
    Human (*Homo sapiens*) 111
    Moose (*Alces alces*) 112
    Elk (*Cervus elaphus*) 112
    Deer (*Odocoileus* sp.) 113
    Bison (*Bison bison*) 113
    Cow (*Bos taurus*) 114
    Antelope (*Antilocapra americana*) 114
    Mountain sheep (*Ovis canadensis*) 115
    Domestic sheep (*Ovis aries*) 115
    Domestic pig (*Sus scrofa*) 116
    Llama (*Lama glama*) 116
    Horse (*Equus*) 117
    Bear (*Ursus americanus*) 117
    Wolf (*Canis lupus*) 118
    Coyote (*Canis latrans*) 118
    Mountain lion (*Felis concolor*) 119
    Bobcat (*Lynx rufus*) 119
    Raccoon (*Procyon lotor*) 120
    Badger (*Taxidea taxus*) 120
    Skunk (*Mephitis mephitis*) 121
    River otter (*Lontra canadensis*) 121
    Rabbit (*Lepus* sp.) 122
    Beaver (*Castor canadensis*) 122
    Porcupine (*Erethizon dorsatum*) 123
    Marmot (*Marmota monax*) 123
    Prairie dog (*Cynomys gunnisoni*) 124
    Norway rat (*Rattus norvegicus*) 124
    Squirrel (*Sciuridae sciurus niger*) 125
    Armadillo (*Dasypus novemcinctus*) 125
    Opossum (*Didelphis virginiana*) 126
    Seal (*Phoca vitulina*) 126

Features of the Humerus

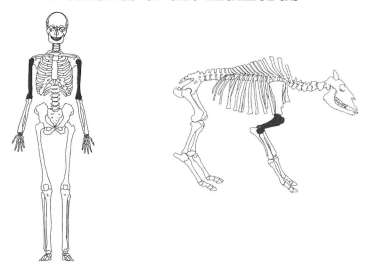

The humerus is the most massive bone of the upper limb or forelimb. The rounded head at the proximal end articulates with the glenoid fossa of the scapula and the distal end articulates with the radius and ulna. All animals have a rounded humeral head, but the areas for insertion of the muscles surrounding the joint vary considerably. At the elbow joint, the trochlea articulates with the ulna and the capitulum articulates with the radius. The distal humerus and the proximal radius and ulna are quite different in the human, large quadrupeds, and smaller quadrupeds. Most large quadrupeds carry more weight over the forelimbs than on the hindlimbs, as the center of gravity is more cranially located.

The humeral head in large quadrupeds faces more posteriorly.

Human have a moderate deltoid tuberosity (the raised area for the insertion of the deltoid muscle which draws the arm away from the midline), but in some animals, such as the beaver or badger (animals with powerful digging habits), the deltoid tuberosity is quite large.

Human and Nonhuman Bone Identification

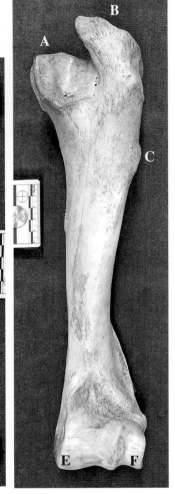
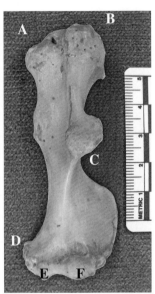

Human   Moose   Beaver

A. Humeral head
B. Greater tubercle
C. Deltoid tuberosity
D. Medial epicondyle
E. Trochlea
F. Capitulum

Humerus

Human (*Homo sapiens*)

Anterior　　Posterior

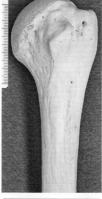
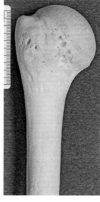
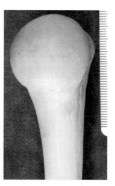

Proximal aspect

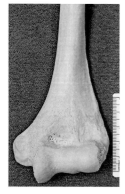
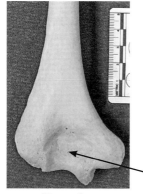

Anterior　　Posterior

Distal aspect

Supratrochlear foramen is variable and depends at least to some degree on the size of the bone. This humerus has no foramen.

Moose (*Alces alces*)

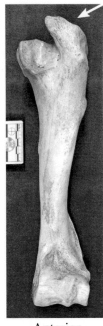
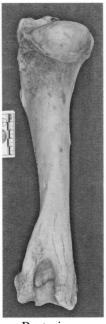

Anterior   Posterior

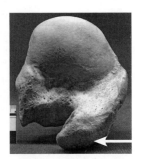
Note large slanted tubercle

Elk (*Cervus elaphus*)

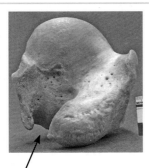
Large, curved tubercle and deep bicipital notch

Note curved tubercle different from moose

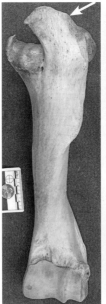
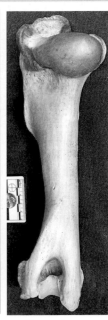

Anterior   Posterior

Humerus

Anterior   Posterior

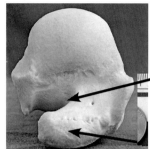

Notch

Large, curved tubercle

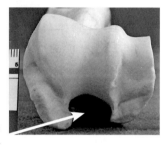

Moderate concavity

Deer (*Odocoileus* sp.)

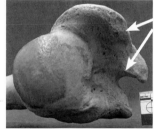

Large, fan-shaped tubercle and notch

Large deltoid tuberosity

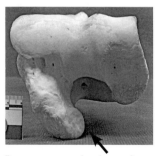

Large, curved tuberosity

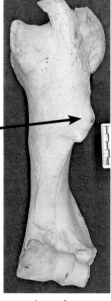
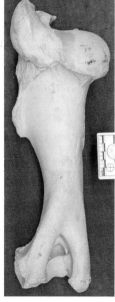

Anterior   Posterior

Bison (*Bison bison*)

Cow (*Bos taurus*)

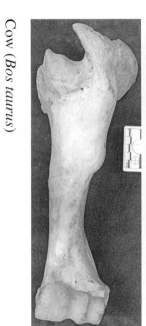
Anterior

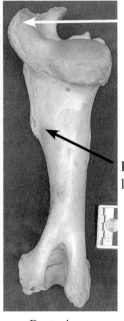
Posterior

Large fan-shaped tubercle

Deltoid tuberosity not as large as in bison

Large curved tuberosity

Antelope (*Antilocapra americana*)

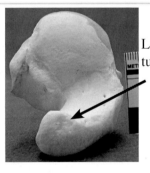
Large, curved tubercle

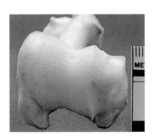
Tuberosity not as pronounced

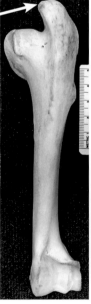
Anterior

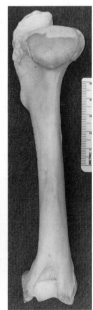
Posterior

Humerus

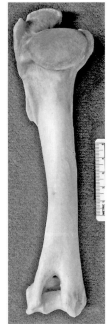

Anterior — Posterior

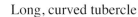
Long, curved tubercle

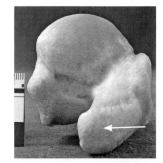

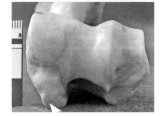

Moderate tuberosity

Mountain sheep (Ovis canadensis)

Very large curved tubercle

Similar to mountain sheep but smaller

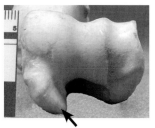

Moderate tuberosity

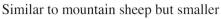

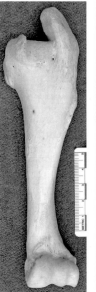

Anterior — Posterior

Domestic sheep (Ovis aries)

Domestic pig (*Sus scrofa*)

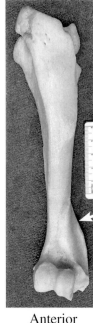
Anterior

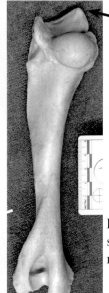
Posterior

Note tubercle shape

Large supinator ridge

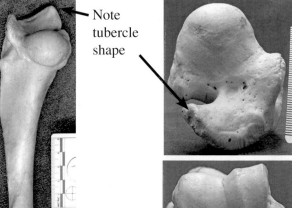

Humerus becomes greater in diameter proximally.

Llama (*Lama glama*)

Lobed tubercle

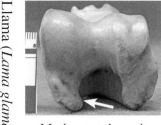
Moderate tuberosity

Low crest

Deltoid tuberosity

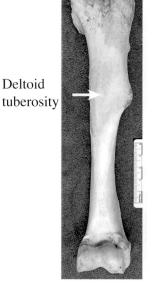
Anterior

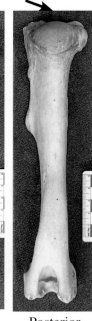
Posterior

Humerus

Horse (*Equus*)

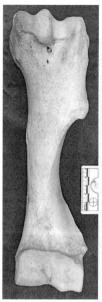
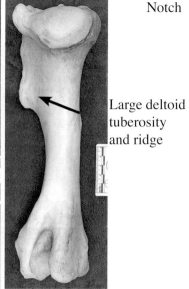

Notch

Large deltoid tuberosity and ridge

Lobed, low tubercle

Large tubercle and deep incurvature

Anterior   Posterior

Bear (*Ursus americanus*)

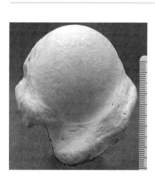

Caution! The bear proximal humerus is similar to human!

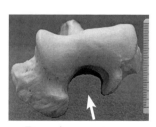

Deep incurvature

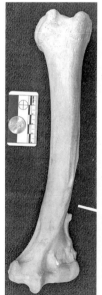
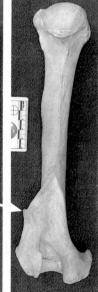

Anterior   Posterior

Fairly well-developed supracondylar crest

118 Human and Nonhuman Bone Identification

Wolf (*Canis lupus*)

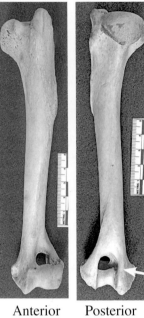
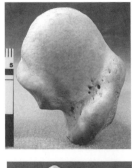

Anterior   Posterior

Large supratrochlear perforation

Coyote (*Canis latrans*)

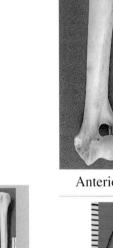
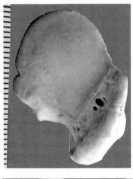
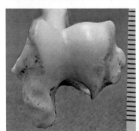

Large supratrochlear perforation

Anterior   Posterior

Humerus

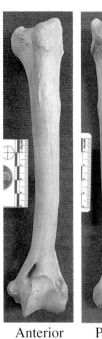

Anterior   Posterior

Entepicondyle foramen

Cats have an entepicondylar foramen and no supratrochlear foramen. Dogs have a supratrochlear foramen and no entepicondylar foramen.

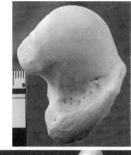

Mountain lion (*Felis concolor*)

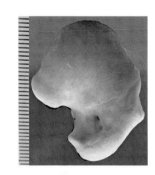
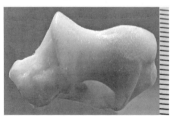

Anterior   Posterior

Bobcat (*Lynx rufus*)

Raccoon (*Procyon lotor*)

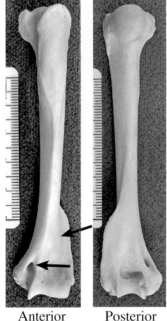

Moderate supracondyloid crest

Small, round entepicondylar foramen

Anterior   Posterior

Badger (*Taxidea taxus*)

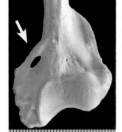

Large deltoid tuberosity

Large supracondylar ridge

Large medial epicondylar ridge

Entepicondylar foramen

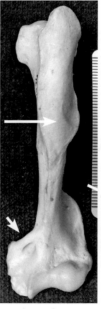
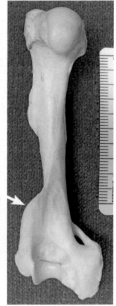

Anterior   Posterior

Humerus

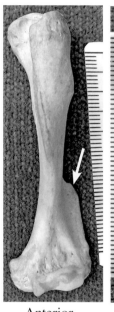
Anterior

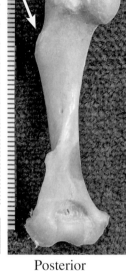
Posterior

Deltoid tuberosity

Supracondyloid crest

No entepicondylar foramen or supracondyloid foramen

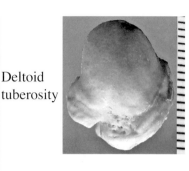

Skunk (*Mephitis mephitis*)

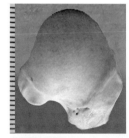

Curved humeral shaft

Supracondyloid ridge

Large medial epicondyle

Anterior

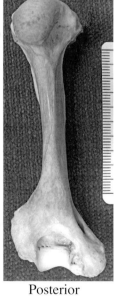
Posterior

River otter (*Lontra canadensis*)

Rabbit (*Lepus* sp.)

Anterior   Posterior

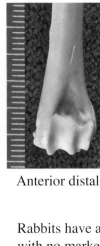

Anterior distal

As in other genera, different species of rabbits exhibit differences in morphology.

Rabbits have a very long, thin shaft with no marked deltoid tuberosity, no pronounced medial epicondyle, and a relatively small tubercle.

Beaver (*Castor canadensis*)

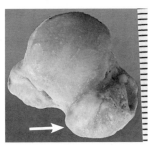

Large tubercle

Pronounced supracondyloid ridge

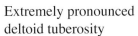

Extremely pronounced deltoid tuberosity

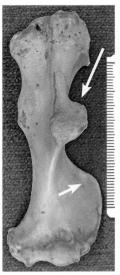 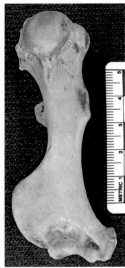

Anterior   Posterior

Humerus

Anterior　　Posterior

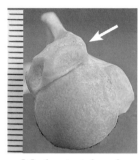

Moderate tubercle

Moderate deltoid tuberosity

Porcupine (Erethizon dorsatum)

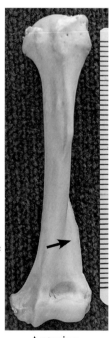

Moderate supracondyloid ridge

No entepicondylar foramen, no supracondyloid foramen

Anterior　　Posterior

Marmot (Marmota monax)

Prairie dog (*Cynomys gunnisoni*)

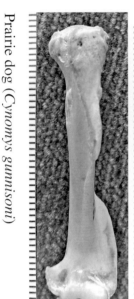
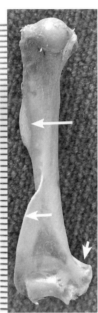

Anterior Posterior

Deltoid tuberosity

Well-developed supracondyloid ridge

Well-developed medial epicondyle

Norway rat (*Rattus norvegicus*)

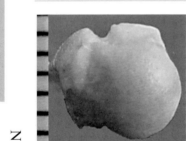
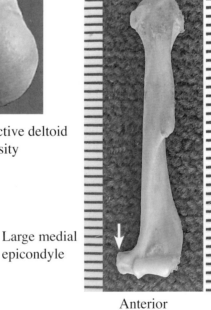
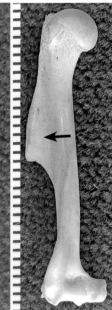

Distinctive deltoid tuberosity

Large medial epicondyle

Anterior Posterior

Humerus

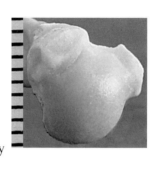

Deltoid tuberosity

Large supracondyloid ridge

Entepicondylar foramen

Anterior　Posterior

Squirrel (*Sciuridae sciurus niger*)

Large deltoid tuberosity and ridge

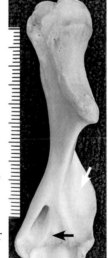
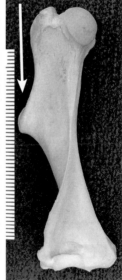

Supracondylar ridge

Entepicondylar foramen

Anterior　Posterior

Armadillo (*Dasypus novemcinctus*)

Opossum (*Didelphis virginiana*)

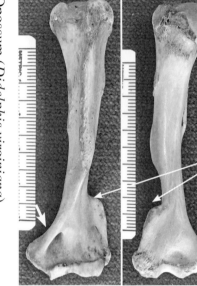

Anterior Posterior

Large supracondylar ridge with abrupt superior angle

Entepicondylar foramen

Seal (*Phoca vitulina*)

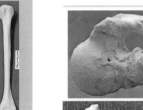

The humerus in a seal widens significantly toward the flipper. Note large muscle attachment ridges near distal aspect.

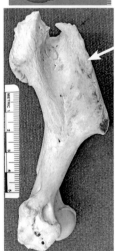
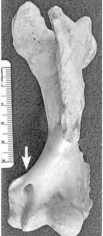
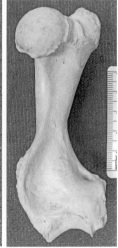

Large deltoid tuberosity and ridge

Entepicondylar foramen

Lateral Anterior Posterior

Radius

Features of the radius 129
    Human (*Homo sapiens*) 131
    Moose (*Alces alces*) 133
    Elk (*Cervus elaphus*) 133
    Deer (*Odocoileus* sp.) 134
    Bison (*Bison bison*) 134
    Cow (*Bos taurus*) 135
    Antelope (*Antilocapra americana*) 135
    Mountain sheep (*Ovis canadensis*) 136
    Domestic sheep (*Ovis aries*) 136
    Domestic pig (*Sus scrofa*) 137
    Llama (*Lama glama*) 137
    Horse (*Equus*) 138
    Bear (*Ursus americanus*) 138
    Wolf (*Canis lupus*) 139
    Coyote (*Canis latrans*) 139
    Mountain lion (*Felis concolor*) 140
    Bobcat (*Lynx rufus*) 140
    Raccoon (*Procyon lotor*) 141
    Badger (*Taxidea taxus*) 141
    Skunk (*Mephitis mephitis*) 142
    River otter (*Lontra canadensis*) 142
    Rabbit (*Lepus* sp.) 143
    Beaver (*Castor canadensis*) 143
    Porcupine (*Erethizon dorsatum*) 144
    Marmot (*Marmota monax*) 144
    Prairie dog (*Cynomys gunnisoni*) 145
    Norway rat (*Rattus norvegicus*) 145
    Squirrel (*Sciuridae sciurus niger*) 146
    Armadillo (*Dasypus novemcinctus*) 146
    Opossum (*Didelphis virginiana*) 147
    Seal (*Phoca vitulina*) 147

Features of the Radius

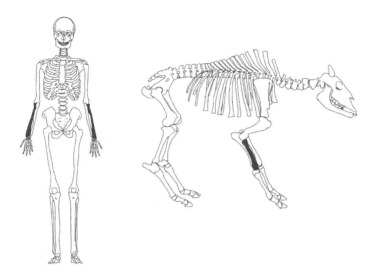

The radius and ulna are the bones of the forearm, and are separate bones in humans. The radius moves over the ulna in pronation in humans and in some other animals, but in many animals the bones are fused and/or the ulna is significantly reduced in size so as to make rotation impossible.

In humans the ulna articulates with the radius distally at the ulnar notch. In those animals in which the ulna is fused or reduced in size, there may be no ulnar notch (arrow points to ulnar notch in humans).

Because the radius and ulna are fused in some animals, some of the following photographs of the radius also show the ulna.

Note some general differences in the photographs (proximal aspects top photograph, distal aspects bottom).

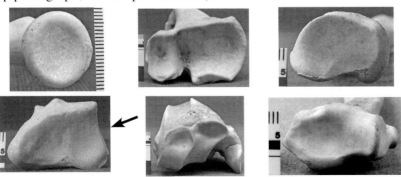

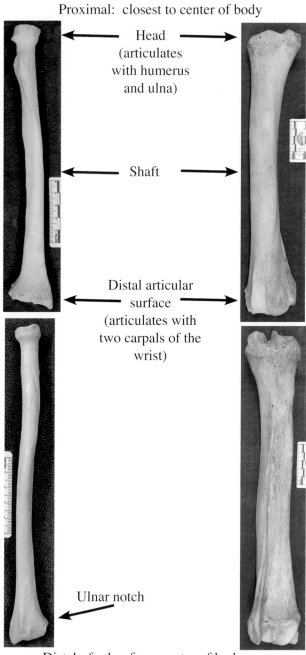

Human and Nonhuman Bone Identification

Proximal: closest to center of body

Head (articulates with humerus and ulna)

Shaft

Distal articular surface (articulates with two carpals of the wrist)

Ulnar notch

Distal: farther from center of body

Human   Moose

Radius

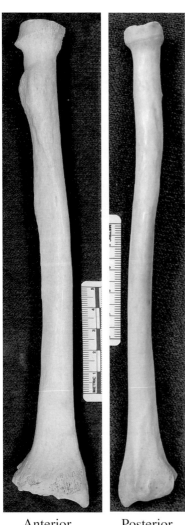

Anterior   Posterior

Proximal aspect

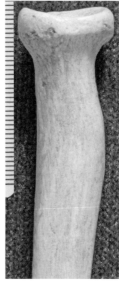

Human (*Homo sapiens*)

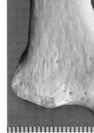

Distal aspect

132   Human and Nonhuman Bone Identification

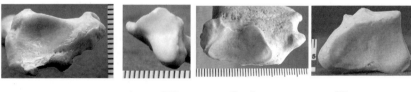

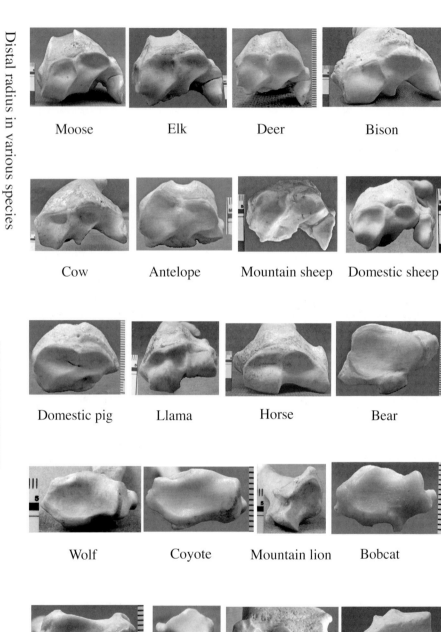

Distal radius in various species

Moose   Elk   Deer   Bison

Cow   Antelope   Mountain sheep   Domestic sheep

Domestic pig   Llama   Horse   Bear

Wolf   Coyote   Mountain lion   Bobcat

Badger   Armadillo   Seal   Human

Radius

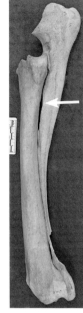

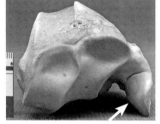

Note gap

Notch on proximal posterior aspect

Radius and ulna are fused in the adult.

Anterior Lateral

Distal aspect Ulna

Moose (Alces alces)

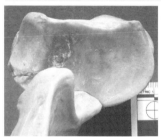
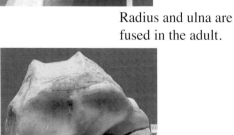

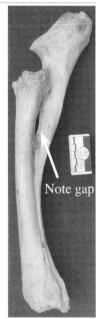

Proximal aspect

Radius and ulna are fused in the adult.

Note gap

Ulna

Distal aspect Anterior Lateral

Elk (Cervus elaphus)

Deer (*Odocoileus* sp.)

Anterior   Posterior

Radius and ulna fused in the adult

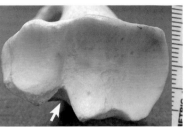

Rounded notch on proximal aspect

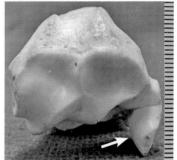

Ulna

Bison (*Bison bison*)

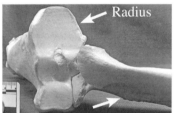

Proximal ulna

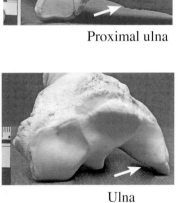

Distal aspect   Ulna

Anterior

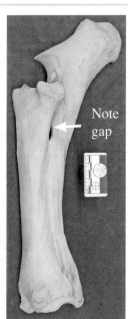

Note gap

Lateral

Radius

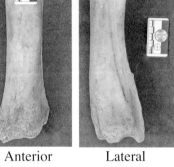

Anterior   Lateral

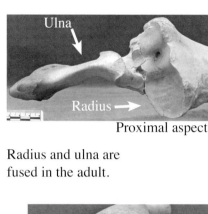

Proximal aspect

Radius and ulna are fused in the adult.

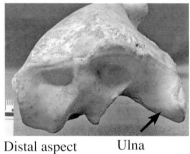

Distal aspect   Ulna

Cow (*Bos taurus*)

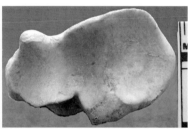

Proximal aspect

Radius and ulna are fused in the adult.

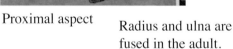

Distal aspect

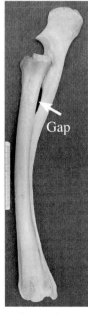

Anterior   Lateral

Antelope (*Antilocapra americana*)

Mountain sheep (*Ovis canadensis*)

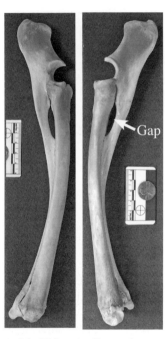

Medial   Lateral

Gap

Proximal aspect

Radius and ulna are fused in the adult.

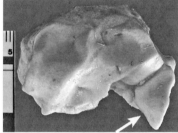

Ulna

Domestic sheep (*Ovis aries*)

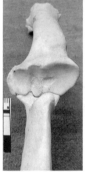

Radius and ulna are fused in the adult.

Smaller gap (arrow)

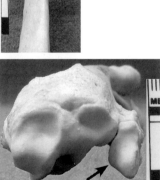

Ulna

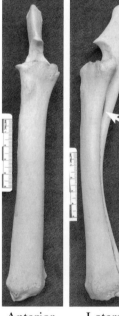

Anterior   Lateral

Radius

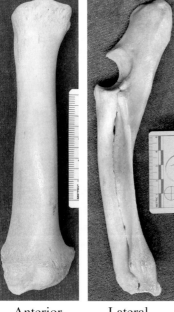

Anterior   Lateral

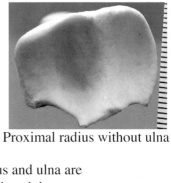

Proximal radius without ulna

The radius and ulna are fused in the adult.

Ulna

Domestic pig (Sus scrofa)

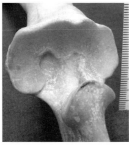

Proximal with ulna

The radius and ulna are fused in the adult and very little of the shaft of the ulna remains.

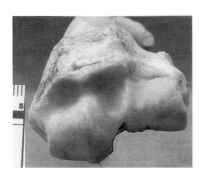

Anterior   Posterior

Llama (Lama glama)

Horse (*Equus*)

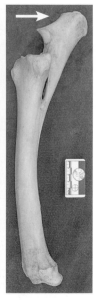

Anterior Posterior

Bulbous olecranon process

Radius and ulna are fused in adults.

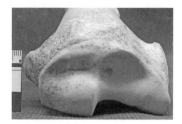

Bear (*Ursus americanus*)

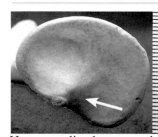

Human radius has smooth border. Bear has projection.

Caution! The bear radius is similar to humans!

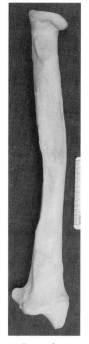

Anterior Posterior

Radius

Anterior Posterior

Proximal

Distal

Wolf (*Canis lupus*)

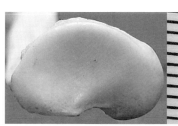

Proximal

Distal

Anterior Posterior

Coyote (*Canis latrans*)

Mountain lion (*Felis concolor*)

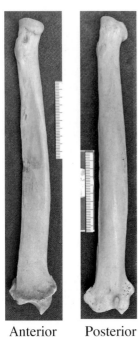

Anterior Posterior

Proximal

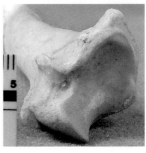

Distal

Bobcat (*Lynx rufus*)

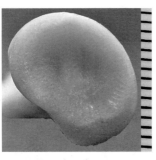

Proximal

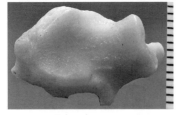

Distal

Anterior Posterior

Radius

Anterior Posterior

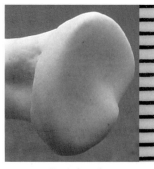

Proximal

Distal

Raccoon (*Procyon lotor*)

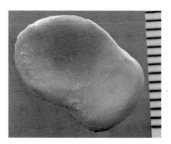

Proximal

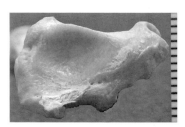

Distal

 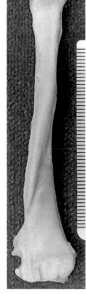

Anterior Posterior

Badger (*Taxidea taxus*)

142　　　　　　　　　　　　　　　　　　　　Human and Nonhuman Bone Identification

Skunk (*Mephitis mephitis*)

Anterior　　Posterior

Proximal

Distal

River otter (*Lontra canadensis*)

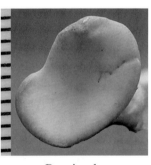

Proximal

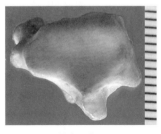

Distal

Anterior　　Posterior

Radius

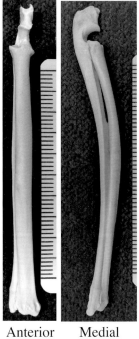

Anterior Medial

Lateral proximal

Lateral distal

Rabbit (*Lepus* sp.)

Proximal

Distal

Anterior Posterior

Beaver (*Castor canadensis*)

Porcupine (*Erethizon dorsatum*)

Anterior

Posterior

Proximal

Distal

Marmot (*Marmota monax*)

Proximal

Distal

Anterior

Posterior

Radius

Anterior   Posterior

Proximal

Distal

Prairie dog (*Cynomys gunnisoni*)

Lateral proximal

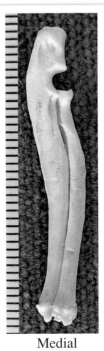 

Medial   Lateral

Norway rat (*Rattus norvegicus*)

146 Human and Nonhuman Bone Identification

Squirrel (*Sciuridae sciurus niger*)

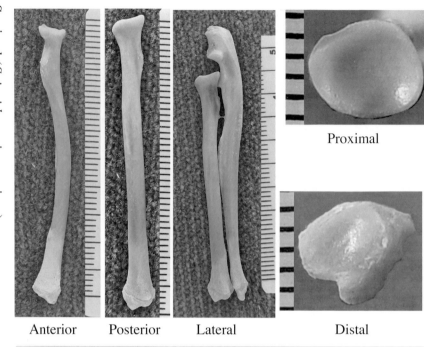

Anterior   Posterior   Lateral   Proximal

Distal

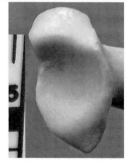

Notice elongated radial head (above) and elongated distal articular surface (below).

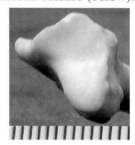

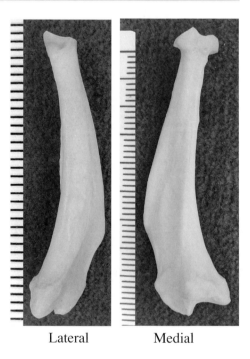

Lateral   Medial

Armadillo (*Dasypus novemcinctus*)

Radius

Anterior

Posterior

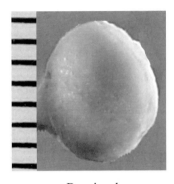
Proximal

Opossum (*Didelphis virginiana*)

Proximal

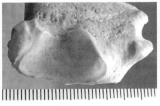
Distal

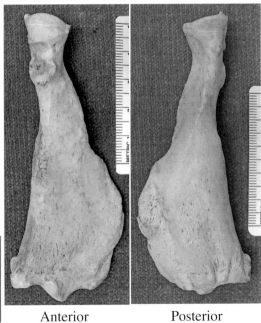
Anterior   Posterior
Notice how the radius widens distally.

Seal (*Phoca vitulina*)

Ulna

Features of the ulna 151
    Human (*Homo sapiens*) 153
    Moose (*Alces alces*) 154
    Elk (*Cervus elaphus*) 154
    Deer (*Odocoileus* sp.) 155
    Bison (*Bison bison*) 155
    Cow (*Bos taurus*) 156
    Antelope (*Antilocapra americana*) 156
    Mountain sheep (*Ovis canadensis*) 157
    Domestic sheep (*Ovis aries*) 157
    Domestic pig (*Sus scrofa*) 158
    Llama (*Lama glama*) 158
    Horse (*Equus*) 159
    Bear (*Ursus americanus*) 159
    Wolf (*Canis lupus*) 160
    Coyote (*Canis latrans*) 160
    Mountain lion (*Felis concolor*) 161
    Bobcat (*Lynx rufus*) 161
    Raccoon (*Procyon lotor*) 162
    Badger (*Taxidea taxus*) 162
    Skunk (*Mephitis mephitis*) 163
    River otter (*Lontra canadensis*) 163
    Rabbit (*Lepus* sp.) 164
    Beaver (*Castor canadensis*) 164
    Porcupine (*Erethizon dorsatum*) 165
    Marmota (*Marmota monax*) 165
    Prairie dog (*Cynomys gunnisoni*) 166
    Norway rat (*Rattus norvegicus*) 166
    Squirrel (*Sciuridae sciurus niger*) 167
    Armadillo (*Dasypus novemcinctus*) 167
    Opossum (*Didelphis virginiana*) 168
    Seal (*Phoca vitulina*) 168

Features of the Ulna

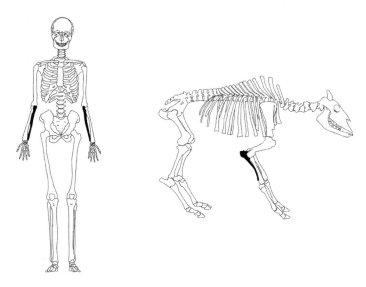

The ulna is the longer bone in the forearm and is responsible for flexion and extension. Because the radius is responsible for rotation, in those animals in which the radius and ulna are fused or in which the ulna is vestigial, no rotation of the forearm is possible.

The radius is lateral to the ulna, although in some animals it is nearly anterior to the ulna.

In many of the photographs of the ulna, the radius is, of necessity, included.

Be aware that the radius and ulna may be separate bones in immature animals even if they are fused in the adult.

 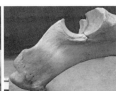 

Human (*Homo sapiens*)   Cow *(Bos taurus)*   Wolf (*Canis lupus*)

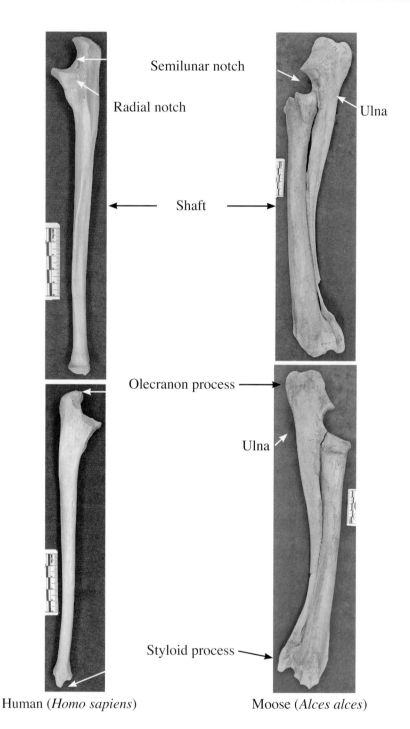

Human (*Homo sapiens*)   Moose (*Alces alces*)

Ulna

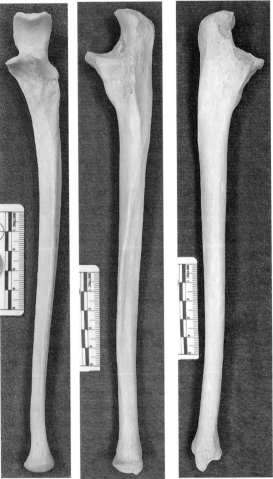
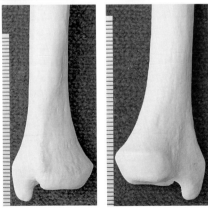

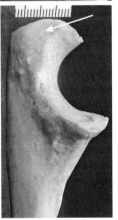

Human (*Homo sapiens*)

Notice very short olecranon process (arrow).

Moose (*Alces alces*)

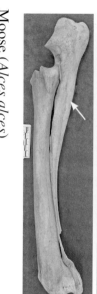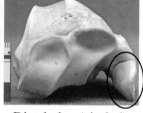

Radius and ulna are fused in the adult.

Arrows point to ulna | Immature radius (left) and ulna | Distal ulna (circled)

Elk (*Cervus elaphus*)

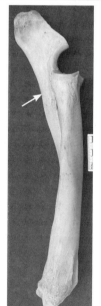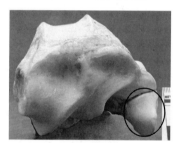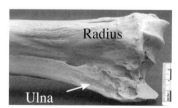

Radius and ulna are fused in the adult.

Distal ulna (circled)

Arrows point to ulna | Distal radius and ulna

Ulna

Distal ulna fused to distal radius.

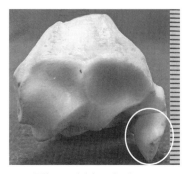

Ulna within circle.

Deer (*Odocoileus* sp.)

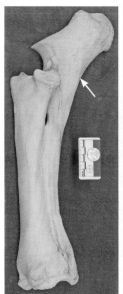
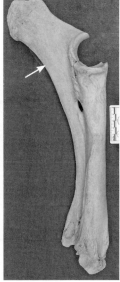

Arrows point to ulna.

Radius and ulna are fused in the adult.

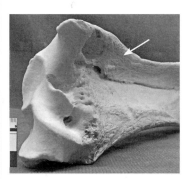

Arrow points to distal ulna fused to radius.

Bison (*Bison bison*)

Cow (*Bos taurus*)

Long olecranon process

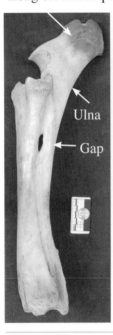
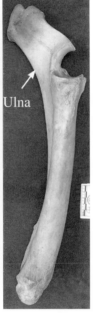

Ulna
Ulna
Gap

Radius and ulna are fused in the adult.

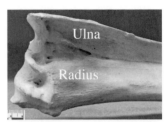

Ulna
Radius

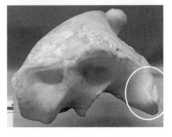

Ulna within circle.

Antelope (*Antilocapra americana*)

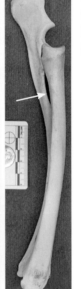

Radius and ulna are fused in the adult. Left three photos show immature individual.

Notice elongated gap between radius and ulna shafts.

Ulna

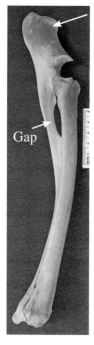

Notice very long olecranon process.

Gap

Radius and ulna are fused in the adult.

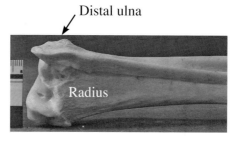

Distal ulna

Radius

Mountain sheep (Ovis canadensis)

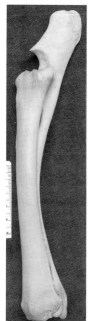
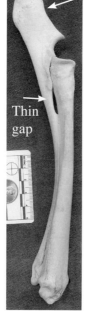

Notice very long olecranon process.

Radius and ulna are fused in the adult.

Thin gap

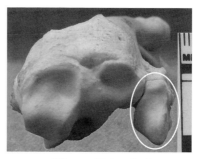

Ulna within circle.

Domestic sheep (Ovis aries)

Domestic pig (*Sus scrofa*)

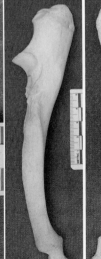
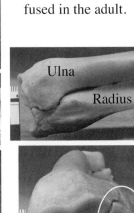

Radius and ulna are fused in the adult.

Arrow points to ulna.

Immature ulna unfused with radius.

Distal ulna within circle.

Llama (*Lama glama*)

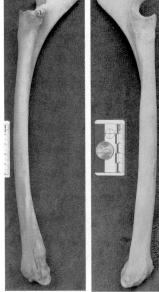

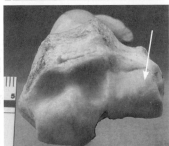

Radius and ulna are fused in the adult.

Arrows point to distal ulna.

Ulna

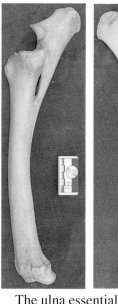

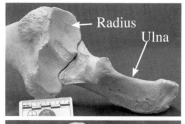
Radius and ulna are fused in the adult.

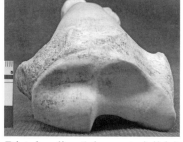

The ulna essentially "melts" into the radius distally.

Distal radius (ulna not visible)

Horse (Equus)

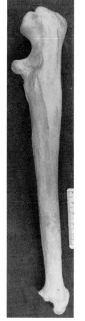

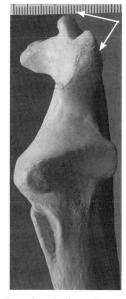

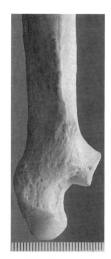

Ulna is broad proximally and tapers distally.

Proximal ulna shown above. Note shape of olecranon process.

Distal ulna

Bear (Ursus americanus)

Wolf (*Canis lupus*)

The olecranon process is lobed, similar to the coyote

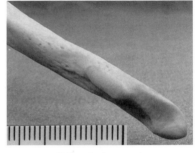

Distal ulna

Coyote (*Canis latrans*)

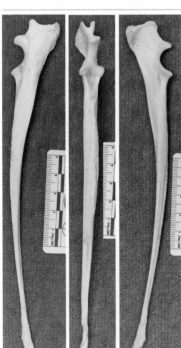

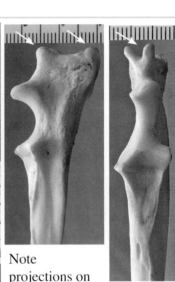

Note projections on olecranon process

Distal ulna

Ulna

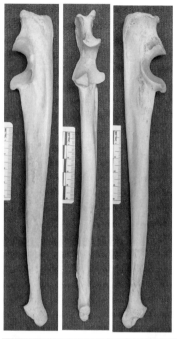

Mountain lion (*Felis concolor*)

Notice bilobed olecranon process
(square in lateral or medial view)

Note knobbed,
square olecranon
process (arrow)

Distal ulna

Bobcat (*Lynx rufus*)

Raccoon (Procyon lotor)

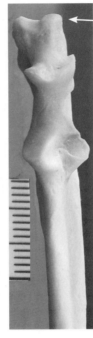

Knobbed olecranon process

Distal ulna

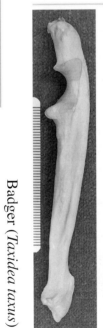

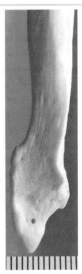

Long, curved olecranon process

Distal ulna

Badger (Taxidea taxus)

Ulna

Note shape of olecranon process

Distal ulna

Skunk (*Mephitis mephitis*)

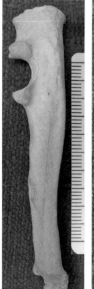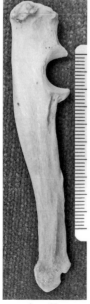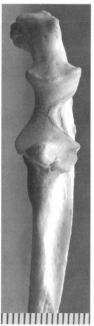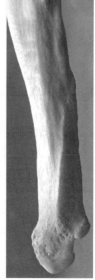

Ulna is wide proximally and tapers distally.

Note shape of olecranon process

Distal ulna

River otter (*Lontra canadensis*)

Rabbit (*Lepus* sp.)

Long, narrow radius and ulna fused in the adult.

Distal radius and ulna. Arrow points to ulna.

Beaver (*Castor canadensis*)

Note shape of olecranon process (arrow).

Distal ulna

Ulna

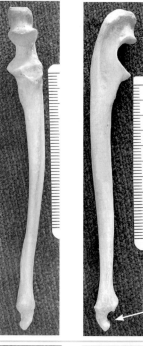

Long styloid process on distal ulna

Porcupine (*Erethizon dorsatum*)

Note shape of olecranon process.

Marmot (*Marmota monax*)

Prairie dog (*Cynomys gunnisoni*)

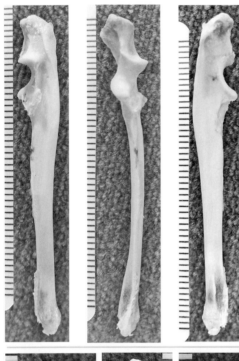

Norway rat (*Rattus norvegicus*)

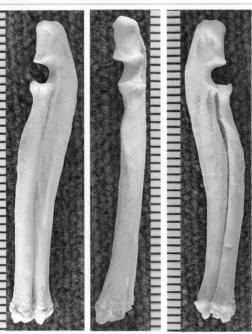

Radius and ulna fused in the adult.

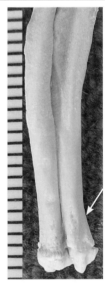

Distal radius and ulna. Arrow points to ulna.

Ulna

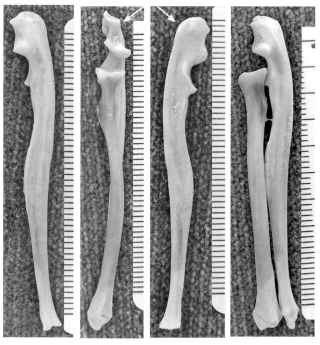

Note shape of olecranon process

Squirrel (*Sciuridae sciurus niger*)

Rounded distal ulna

Very long olecranon process

Armadillo (*Dasypus novemcinctus*)

Opossum (*Didelphis virginiana*)

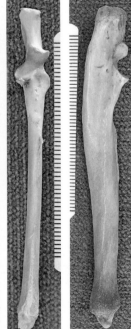

Proximal ulna

Distal ulna

Seal (*Phoca vitulina*)

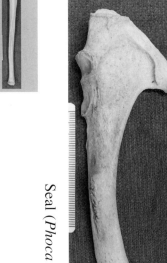

Hatchet-shaped ulna

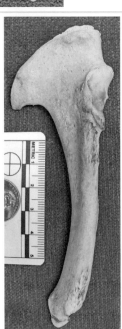

Proximal ulna

Metacarpals and Forelimbs

Features of the metacarpals and forelimbs 171
    Human (*Homo sapiens*) 175
    Moose (*Alces alces*) 176
    Elk (*Cervus elaphus*) 176
    Deer (*Odocoileus* sp.) 177
    Bison (*Bison bison*) 177
    Cow (*Bos taurus*) 178
    Antelope (*Antilocapra americana*) 178
    Mountain sheep (*Ovis canadensis*) 179
    Domestic sheep (*Ovis aries*) 179
    Domestic pig (*Sus scrofa*) 180
    Llama (*Lama glama*) 180
    Horse (*Equus*) 181
    Bear (*Ursus americanus*) 181
    Wolf (*Canis lupus*) 182
    Mountain lion (*Felis concolor*) 182
    Raccoon (*Procyon lotor*) 183
    Badger (*Taxidea taxus*) 183
    Skunk (*Mephitis mephitis*) 184
    Rabbit (*Lepus* sp.) 184
    Marmot (*Marmota monax*) 185
    Norway rat (*Rattus norvegicus*) 185
    Vampire bat (*Vampyressa nymphaea*) 186
    Seal (*Phoca vitulina*) 186

Features of the Metacarpals and Forelimbs

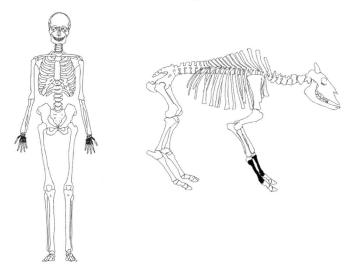

Note: Some of the basic information presented in this section is either repeated or analogous to that presented in the Metatarsal section.

The metacarpals are the bones that make up the palm in humans and part of the leg (in large animals) or paw (in smaller animals). The number and morphology of metacarpals in humans and nonhumans have everything to do with how the hands, forepaws, and forelimbs are used (locomotion, manipulation, etc.)

An array of five digits (for the hand or paw) is the most primitive condition. Many large animals have a reduced number to two (for example in the elk, cow, sheep, etc.) and to one (as in the horse).

In humans, the metacarpals and phalanges are arranged in such a way as to facilitate manipulation of objects, while in many of the nonhuman small animals, the arrangement of the bones of the paw is more appropriate for running or digging. Note on the following page that even the alignment of the metacarpals and phalanges is different depending upon the kind of work the forelimb does. In many animals with paws the metacarpals and phalanges are very tightly grouped, while in humans and in other animals in which manipulation of objects is important, the array of the hand (or paw) is more widely spaced. This information can be important when trying to determine human from nonhuman, particularly in tissued specimens or in an x-ray.

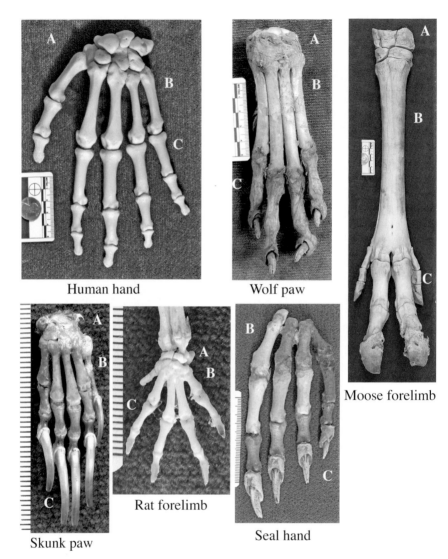

Human hand
Wolf paw
Moose forelimb
Skunk paw
Rat forelimb
Seal hand

Note that there are three rows of phalanges: proximal, middle, and distal.

Distal phalanges are usually claws in small mammals and hooves in large mammals.

        A. Carpals
        B. Metacarpals
        C. Phalanges

Metacarpals and Forelimbs 173

Bear forepaw radiograph. One key area for distinguishing from human is circled.

Caution! Bear paws (forepaws or hindpaws) are often left or thrown away when a bear is skinned. Without the claws (distal phalanges) and particularly if the paw still has soft tissue attached (as in the above left photograph), the paw looks very much like a human hand or foot.

There are several ways to determine whether or not the item is human.

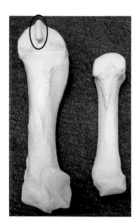 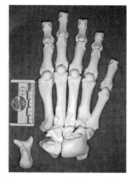 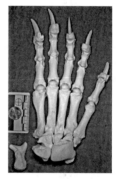

Bear paw without claws   Bear paw with claws

Bear metacarpal (left) and human (right). Area circled on bear shows ridge at center of distal articular surface, also circled on radiograph (above).

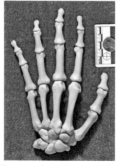 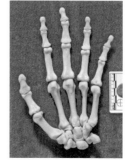

Human hand posterior   Human hand palmar

Human and Nonhuman Bone Identification

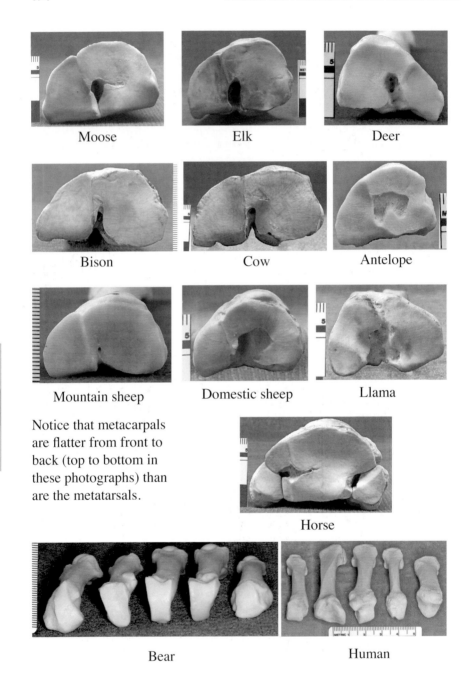

Notice that metacarpals are flatter from front to back (top to bottom in these photographs) than are the metatarsals.

Miscellaneous proximal metacarpals

Metacarpals and Forelimbs

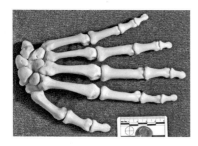

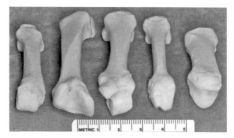

Metacarpals proximal view

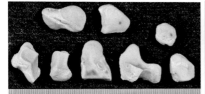

Carpals

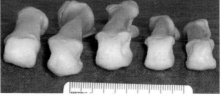

Metacarpals distal view

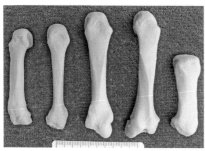

Metacarpals posterior

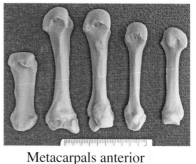

Metacarpals anterior

Hand phalanges

Proximal phalanges, proximal view

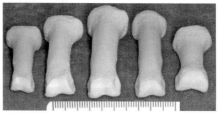

Proximal phalanges, distal view

Human (*Homo sapiens*)

176   Human and Nonhuman Bone Identification

Moose (*Alces alces*)

Anterior   Posterior

Proximal

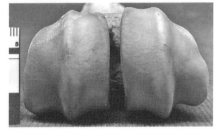

Distal

Elk (*Cervus elaphus*)

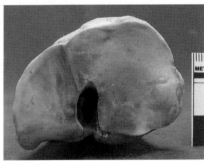

Proximal

Distal

Anterior   Posterior

Metacarpals and Forelimbs

Anterior Posterior

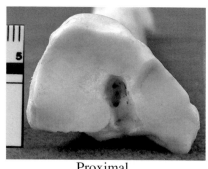

Proximal

Distal

Deer (*Odocoileus* sp.)

Proximal

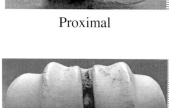

Distal

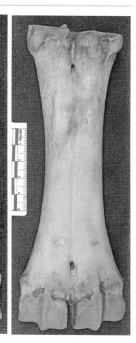

Anterior Posterior

Bison (*Bison bison*)

Cow (*Bos taurus*)

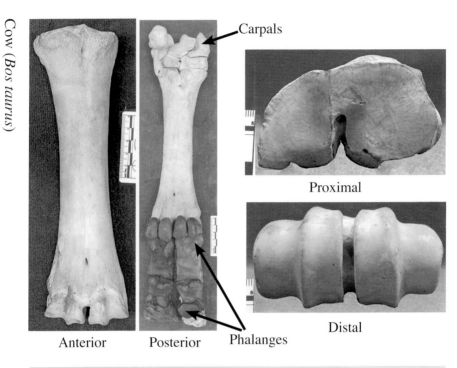

Anterior Posterior Phalanges Proximal Distal Carpals

Antelope (*Antilocapra americana*)

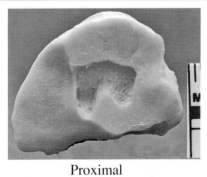

Proximal

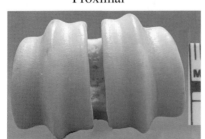

Distal

Anterior Posterior

Metacarpals and Forelimbs

Proximal

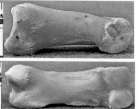
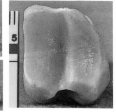

Phalanges

Anterior   Posterior

Mountain sheep (Ovis canadensis)

Proximal

Anterior   Posterior

Domestic sheep (Ovis aries)

Domestic pig (*Sus scrofa*)

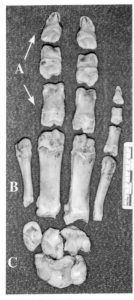

Anterior

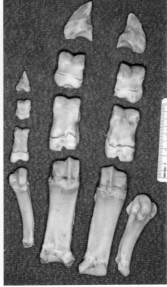

Posterior

A: Phalanges   B: Metacarpals   C: Carpals

Proximal phalanx

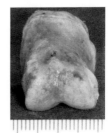

Distal phalanx

Proximal

Llama (*Lama glama*)

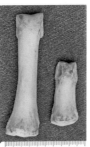

Phalanges

Anterior

Posterior

Metacarpals and Forelimbs

Anterior

Posterior

Proximal metacarpal

Proximal, middle, and distal phalanges shown at right.

Proximal aspect of proximal phalanx

Horse (*Equus*)

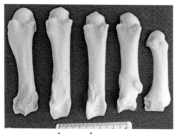

Bear forepaw without claws

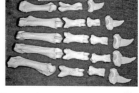

Bear forepaw posterior

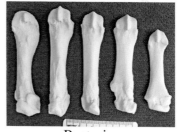

Anterior   Posterior

Proximal

Distal

Bear (*Ursus americanus*)

Wolf (*Canis lupus*)

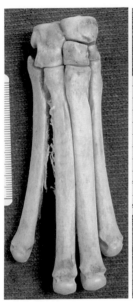
Anterior

Posterior

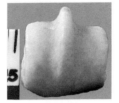
Distal

Mountain lion (*Felis concolor*)

Posterior

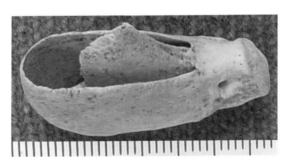

Distal phalanges

Metacarpals and Forelimbs

Carpals

Metacarpals

Phalanges

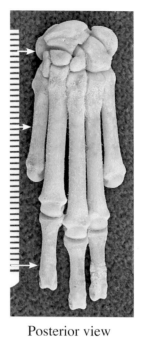

Raccoon (*Procyon lotor*)

Distal phalanges

Posterior view

Radius and ulna

Carpals

Metacarpals

Phalanges

Badger (*Taxidea taxus*)

Notice large claws

Plantar view

Skunk (*Mephitis mephitis*)

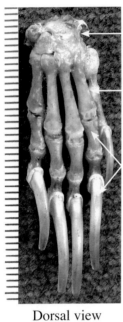

Carpals
Metacarpals
Phalanges

Dorsal view

Rabbit (*Lepus* sp.)

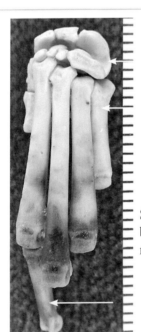
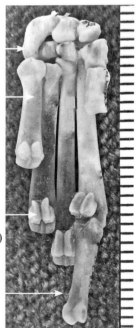

Carpals
Metacarpals
Sesamoid bones (in many animals)
Phalanges

Dorsal view   Palmar view

Metacarpals and Forelimbs

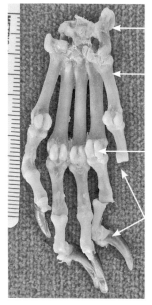

- Carpals
- Metacarpals
- Sesamoid bones (in many animals)
- Phalanges

Palmar view

Marmot (*Marmota monax*)

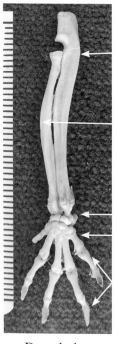

- Ulna
- Radius
- Carpals
- Metacarpals
- Phalanges

Dorsal view

Norway rat (*Rattus norvegicus*)

Vampire bat (*Vampyressa nymphaea*)

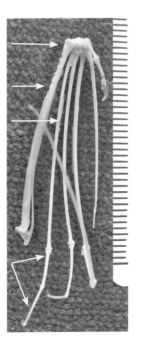

Seal (*Phoca vitulina*)

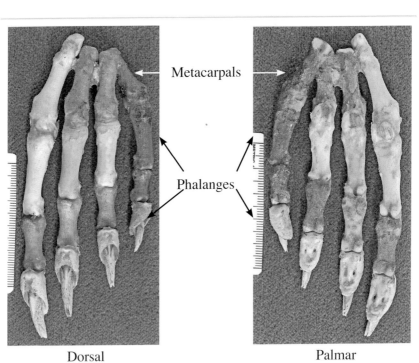

Dorsal   Palmar

Pelvic Girdle

Features of the pelvic girdle 189
    Human (*Homo sapiens*) 191
    Moose (*Alces alces*) 192
    Elk (*Cervus elaphus*) 192
    Deer (*Odocoileus* sp.) 193
    Bison (*Bison bison*) 193
    Cow (*Bos taurus*) 194
    Antelope (*Antilocapra americana*) 194
    Mountain sheep (*Ovis canadensis*) 195
    Domestic sheep (*Ovis aries*) 195
    Domestic pig (*Sus scrofa*) 196
    Llama (*Lama glama*) 196
    Horse (*Equus*) 197
    Bear (*Ursus americanus*) 197
    Wolf (*Canis lupus*) 198
    Coyote (*Canis latrans*) 198
    Mountain lion (*Felis concolor*) 199
    Bobcat (*Lynx rufus*) 199
    Raccoon (*Procyon lotor*) 200
    Badger (*Taxidea taxus*) 200
    Skunk (*Mephitis mephitis*) 201
    River otter (*Lontra canadensis*) 201
    Rabbit (*Lepus* sp.) 202
    Beaver (*Castor canadensis*) 202
    Porcupine (*Erethizon dorsatum*) 203
    Marmota (*Marmota monax*) 203
    Prairie dog (*Cynomys gunnisoni*) 204
    Norway rat (*Rattus norvegicus*) 204
    Squirrel (*Sciuridae sciurus niger*) 205
    Armadillo (*Dasypus novemcinctus*) 205
    Opossum (*Didelphis virginiana*) 206
    Seal (*Phoca vitulina*) 206

Features of the Pelvic Girdle

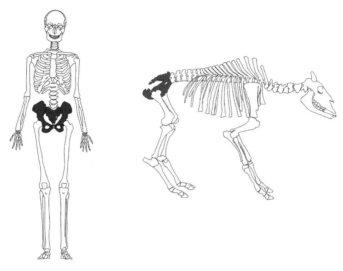

The pelvis in all mammals is comprised of a left and right os coxa (also called innominates) and the sacrum. In many mammals these three elements fuse into one unit in the adult, so they are considered together in this section of the book. Also, the os coxa is made of three bones: the ilium, ischium, and pubis, which fuse into the os coxa (one unit) in the adult.

The sacrum is usually made of 4 to 6 individual sacral vertebrae that fuse in the adult to create a single bone. The area of contact between the last lumbar vertebra and the first sacral vertebra is called the promontory. The wings on either side are the sacral alae.

Some of the animals in this section have had the sacrum fused to the os coxae, a common condition in adult of many of the nonhuman species.

Caution! The femur can be confused with the humerus because of the ball joint at the proximal end. The head of the femur is a more complete ball than is the head of the humerus as it is nestled inside the acetabulum of the os coxa.

Human (*Homo sapiens*)

Antelope (*Antilocapra americana*)

Human and Nonhuman Bone Identification

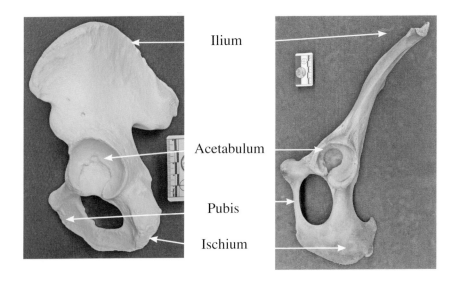

Above: Os coxa of human and bison
Below: Sacrum of human and bison

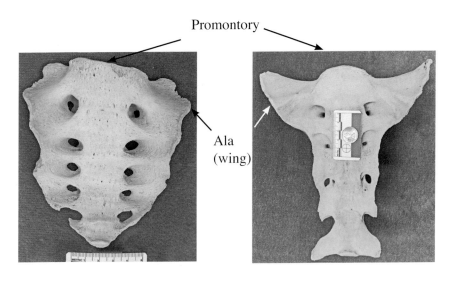

Human   Bison

Pelvic Girdle

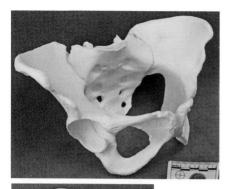

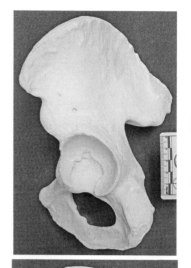

Human (*Homo sapiens*)

Sacrum, anterior view

Wide ilium

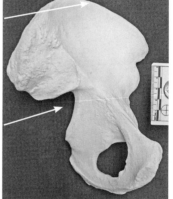

Acute greater sciatic notch

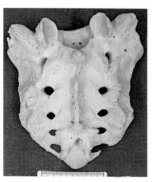

Sacrum, posterior view

Os coxa lateral (above), medial (below)

Sacrum, cranial view

1st coccygeal vertebra

Moose (*Alces alces*)

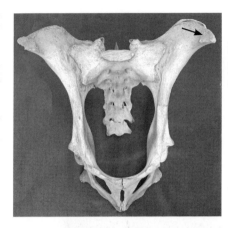
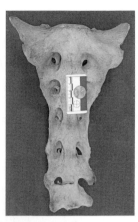

Iliac crests project and curve

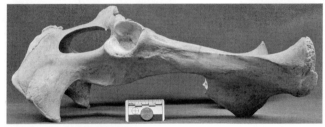

Elk (*Cervus elaphus*)

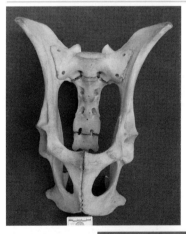
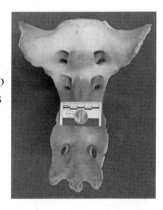

Narrower pelvic girdle than moose. Iliac crests do not project as far.

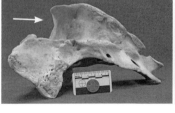
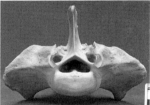

"Blending" of spinous process of sacrum

Pelvic Girdle

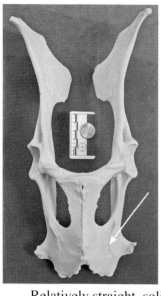
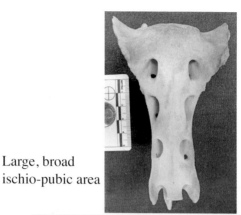

Large, broad ischio-pubic area

Relatively straight, solid spinous process area.

Deer (*Odocoileus* sp.)

Note shape of ilium

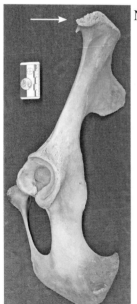
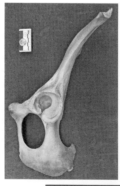
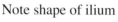

Notice shape of sacrum and spinous process area

Bison (*Bison bison*)

Cow (Bos taurus)

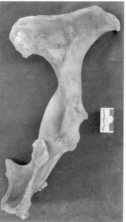

Notice shape of ilium.

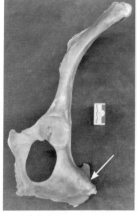

Notice shape of ischium.

This proximal sacrum is pathological.

Note shape of spinous process area.

Antelope (Antilocapra americana)

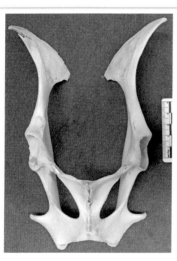

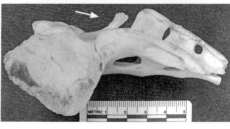

Note shape of spinous process area.

Pelvic Girdle

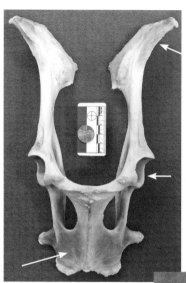

← Notice shape of ilium.

← Notice shape of acetabulum.

Notice shape of ischiopubic area.

Mountain sheep (*Ovis canadensis*)

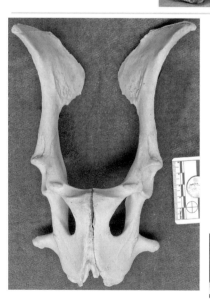

Note shape of spinous process area.

Domestic sheep (*Ovis aries*)

Human and Nonhuman Bone Identification

Domestic pig (*Sus scrofa*)

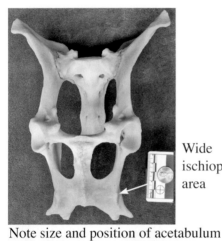

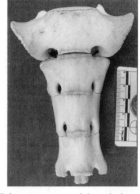

Wide ischiopubic area

Thin sacrum with minimal spinous processes

Note size and position of acetabulum

Wide, squared ilium

Llama (*Lama glama*)

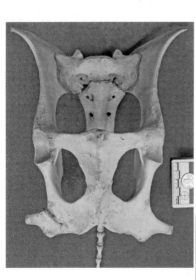

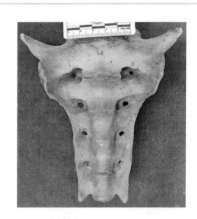

Pelvic Girdle

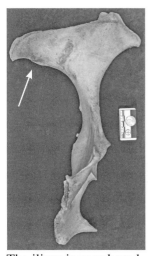
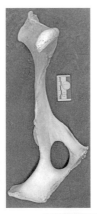
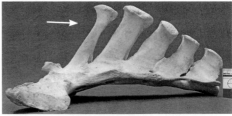

Note shape of sacrum and spinous processes.

The ilium is very broad and does not extend a great distance craniocaudally.

Horse (*Equus*)

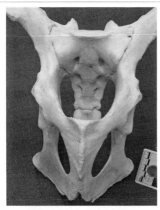
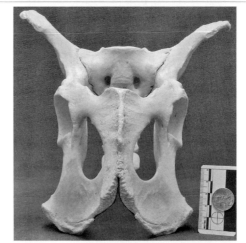

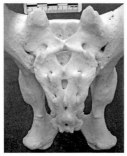

Bear (*Ursus americanus*)

Wolf (*Canis lupus*)

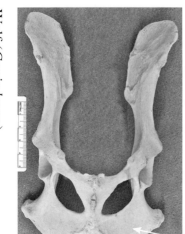

Wider ischiopubic area than in cats.

Coyote (*Canis latrans*)

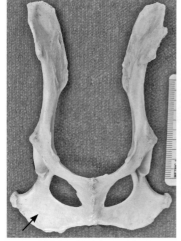

Wider ischiopubic area than in cats.

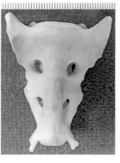

Spinous processes do not project greatly.

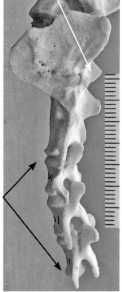

Tail (coccygeal) vertebrae

Pelvic Girdle

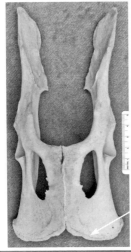

Note shape Long spinous processes

Mountain lion (*Felis concolor*)

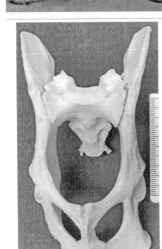

Pelvic girdle with sacrum in place

Note shape of spinous processes

Bobcat (*Lynx rufus*)

Raccoon (*Procyon lotor*)

Note shape

Badger (*Taxidea taxus*)

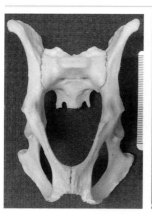
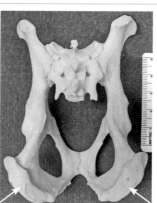
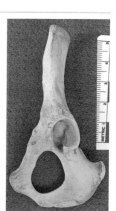

Strong ischial tuberosity

Pelvic Girdle

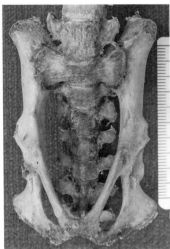

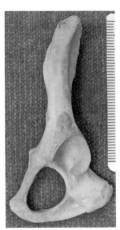

Tail vertebrae

Skunk (*Mephitis mephitis*)

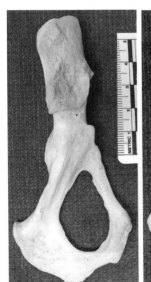
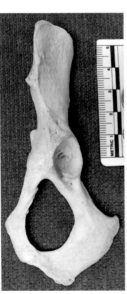

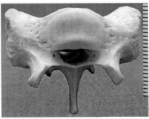

River otter (*Lontra canadensis*)

Human and Nonhuman Bone Identification

Rabbit (*Lepus* sp.)

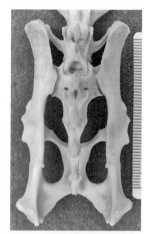

Beaver (*Castor canadensis*)

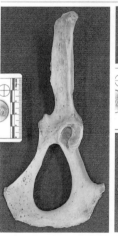
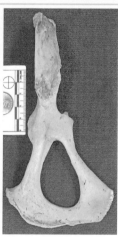
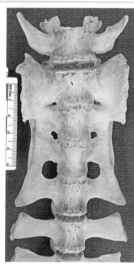

Note shape of spinous processes.

The sacrum of the beaver reflects the large tail. Note that it becomes larger caudally.

Pelvic Girdle

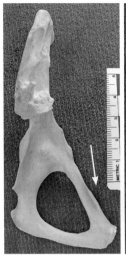

Note shape of sacrum and spinous processes.

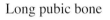

Long pubic bone   Rounded ilium

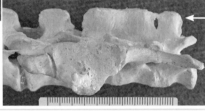

Porcupine (Erethizon dorsatum)

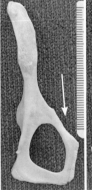
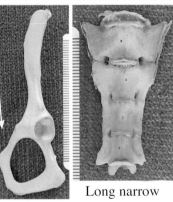

Short pubic bone   Long narrow sacrum

Marmot (Marmota monax)

Prairie dog (*Cynomys gunnisoni*)

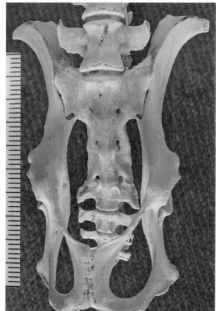
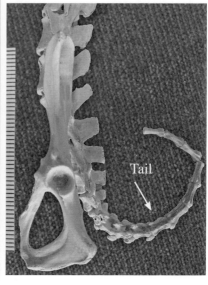

Norway rat (*Rattus norvegicus*)

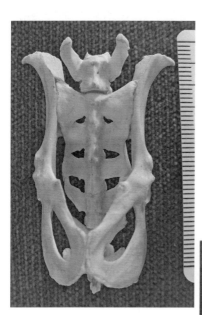
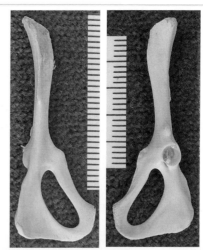

Pelvic Girdle

Squirrel (*Sciuridae sciurus niger*)

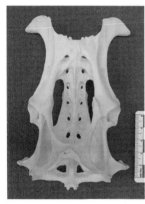
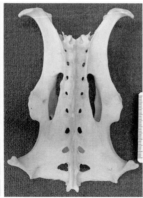

Sacrum widens near tail.

Armadillo (*Dasypus novemcinctus*)

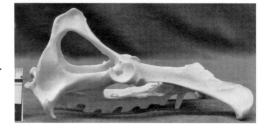

Unusual shape of pelvic girdle.

Opossum (*Didelphis virginiana*)

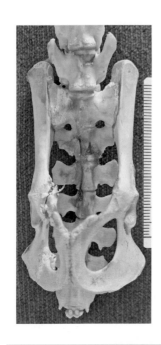
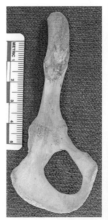
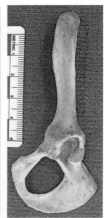

Seal (*Phoca vitulina*)

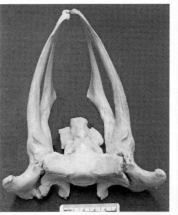
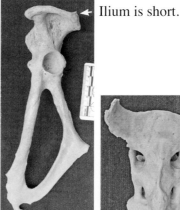

Ilium is short.

Pubis is very long.

Femur

Features of the femur 209
   Human (*Homo sapiens*) 211
   Moose (*Alces alces*) 212
   Elk (*Cervus elaphus*) 212
   Deer (*Odocoileus* sp.) 213
   Bison (*Bison bison*) 213
   Cow (*Bos taurus*) 214
   Antelope (*Antilocapra americana*) 214
   Mountain sheep (*Ovis canadensis*) 215
   Domestic sheep (*Ovis aries*) 215
   Domestic pig (*Sus scrofa*) 216
   Llama (*Lama glama*) 216
   Horse (*Equus*) 217
   Bear (*Ursus americanus*) 217
   Wolf (*Canis lupus*) 218
   Coyote (*Canis latrans*) 218
   Mountain lion (*Felis concolor*) 219
   Bobcat (*Lynx rufus*) 219
   Raccoon (*Procyon lotor*) 220
   Badger (*Taxidea taxus*) 220
   Skunk (*Mephitis mephitis*) 221
   River otter (*Lontra canadensis*) 221
   Rabbit (*Lepus* sp.) 222
   Beaver (*Castor canadensis*) 222
   Porcupine (*Erethizon dorsatum*) 223
   Marmot (*Marmota monax*) 223
   Prairie dog (*Cynomys gunnisoni*) 224
   Norway rat (*Rattus norvegicus*) 224
   Squirrel (*Sciuridae sciurus niger*) 225
   Armadillo (*Dasypus novemcinctus*) 225
   Opossum (*Didelphis virginiana*) 226
   Seal (*Phoca vitulina*) 226

Features of the Femur

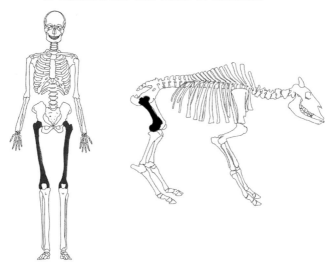

In humans the femur is the longest bone of the body and is long relative to its diameter (though many of the smaller nonhuman mammals also have long femoral shafts relative to the diameter). The head of the femur is round in all animals and articulates with the os coxa (hip joint) at the acetabulum. The articular surface at the distal end (at the knee) is divided into two condyles. These condyles are relatively smooth in all animals, but the anterior articular surface in nonhumans is usually more "sculpted" than in humans.

Caution! The femur can be confused with the humerus because of the ball joint at the proximal end. The head of the femur is a more complete ball than is the head of the humerus as it is nestled inside the acetabulum of the os coxa.

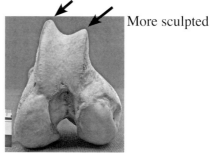

Smoother More sculpted

Human (*Homo sapiens*)   Moose (*Alces alces*)

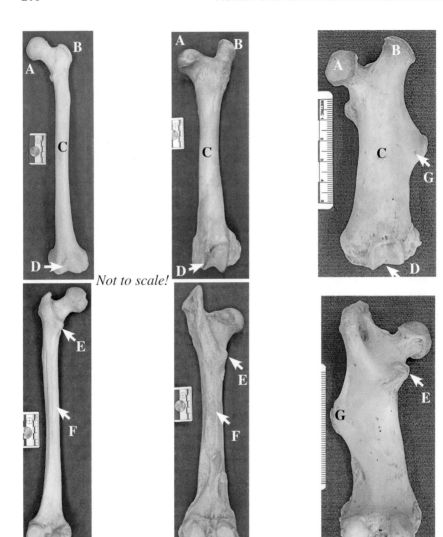

Human | Moose | Beaver

A. Femoral head (articulates with acetabulum in pelvis)
B. Greater Trochanter
C. Shaft
D. Condyles (articulate with tibia)
E. Lesser trochanter
F. Linea aspera (muscle insertion)
G. Third trochanter (not all species)

Femur

Human (*Homo sapiens*)

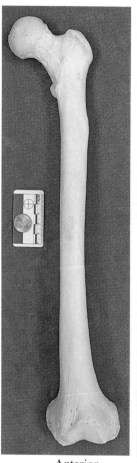
Anterior

Posterior

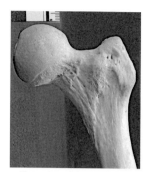
Proximal anterior

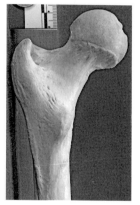
Proximal posterior

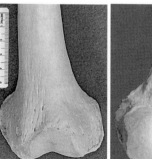
Anterior

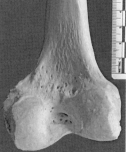
Posterior

Distal aspect

Distal

Moose (*Alces alces*)

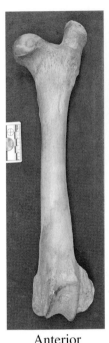
Anterior

Posterior

Large greater trochanter

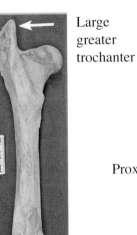
Proximal

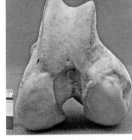
Distal

Elk (*Cervus elaphus*)

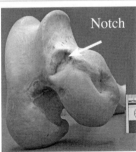
Notch

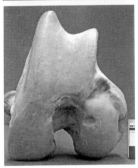
Distal views

Anterior

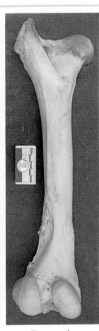
Posterior

Femur

Anterior

Note angle
Posterior

Proximal

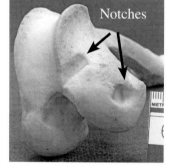
Notches
Distal

Deer (Odocoileus sp.)

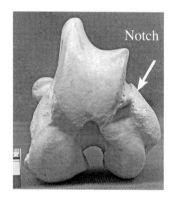
Notch

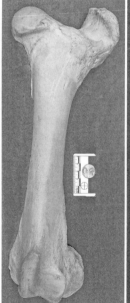
Anterior

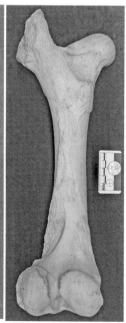
Posterior

Bison (Bison bison)

Cow (*Bos taurus*)

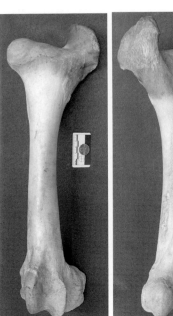
Anterior

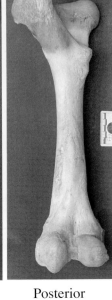
Posterior

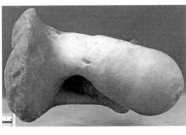
Proximal

Distal

Notch

Antelope (*Antilocapra americana*)

Proximal

Relatively large greater trochanter

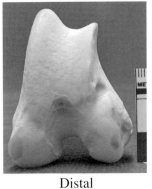
Distal

Relatively slender, long shaft with small lesser trochanter

Anterior

Posterior

Femur

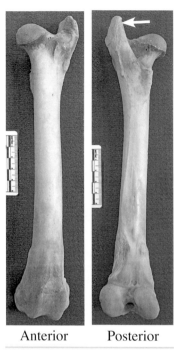

Anterior — Posterior

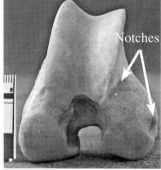

Large greater trochanter

Proximal

Notches

Distal

Mountain sheep (Ovis canadensis)

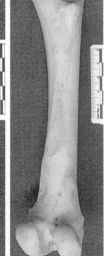

Notches

Anterior — Posterior

Domestic sheep (Ovis aries)

Domestic pig (*Sus scrofa*)

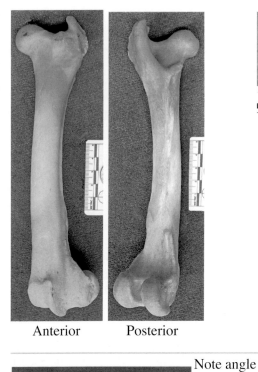

Anterior   Posterior

Small femoral head

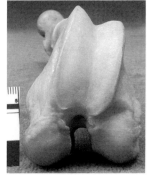

Distal

Proximal

Note angle

Llama (*Lama glama*)

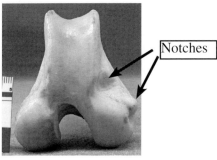

Notches

Distal

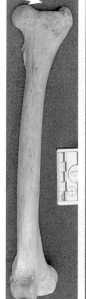 

Anterior   Posterior

Femur

Horse (*Equus*)

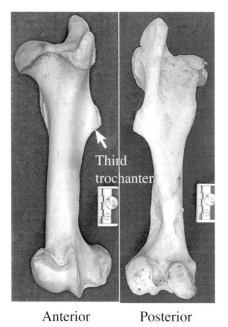
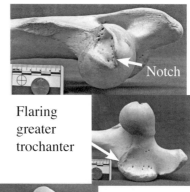

Third trochanter

Notch

Flaring greater trochanter

Anterior   Posterior   Distal

Relatively small greater trochanter

Proximal

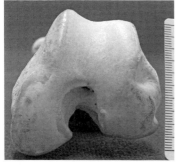

Caution! The distal bear femur is somewhat similar to humans.

Anterior   Posterior

Bear (*Ursus americanus*)

Wolf (*Canis lupus*)

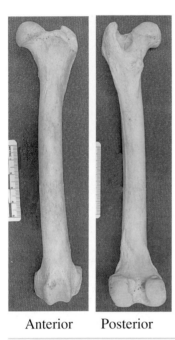

Anterior Posterior

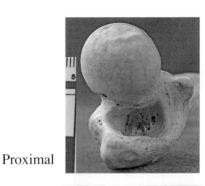

Proximal

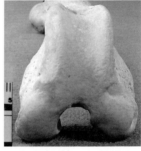

Distal

Coyote (*Canis latrans*)

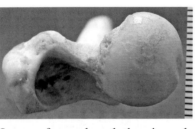

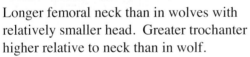

Longer femoral neck than in wolves with relatively smaller head. Greater trochanter higher relative to neck than in wolf.

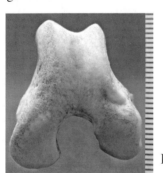

Distal

Anterior Posterior

Femur

Anterior | Posterior

Small greater trochanter, small femoral head

Distal aspects

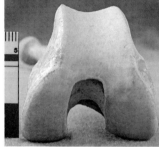

Mountain lion (*Felis concolor*)

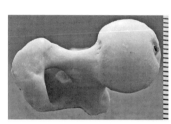

Proximal

Similar to mountain lion, but smaller

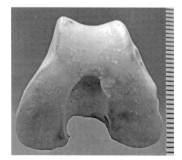

Distal

Anterior | Posterior

Bobcat (*Lynx rufus*)

Raccoon (*Procyon lotor*)

Anterior

Posterior

Small greater trochanter

Small lesser trochanter

Proximal

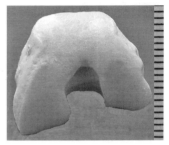
Distal

Badger (*Taxidea taxus*)

Notice ridge below greater trochanter

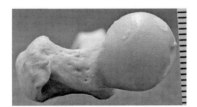
Proximal

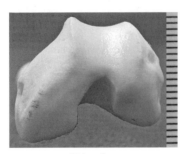
Distal

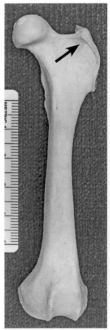
Anterior

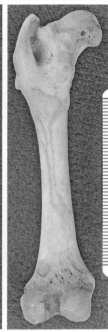
Posterior

Femur

Anterior   Posterior

Proximal

Distal

Skunk (*Mephitis mephitis*)

Proximal

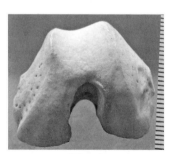

Distal

Femur widens distally ➤

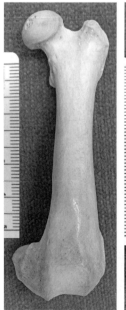 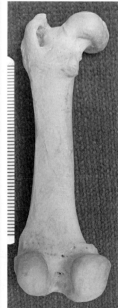

Anterior   Posterior

River otter (*Lontra canadensis*)

Rabbit (*Lepus* sp.)

Anterior   Posterior

Proximal

Third trochanter high on shaft

Distal

Long, narrow femur

Beaver (*Castor canadensis*)

Proximal

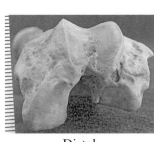

Distal

Very large greater trochanter   Large third trochanter

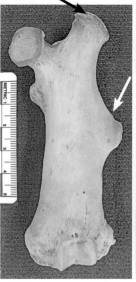 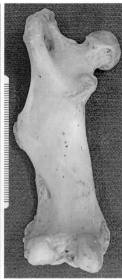

Anterior   Posterior

Femur

Anterior　　　Posterior

Proximal

Relatively broad femur, no third trochanter

Distal

Porcupine (Erethizon dorsatum)

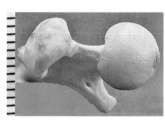

Proximal

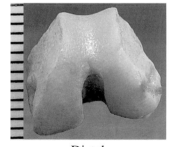

Distal

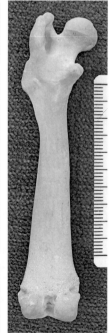

Anterior　　　Posterior

Marmot (Marmota monax)

Prairie dog (*Cynomys gunnisoni*)

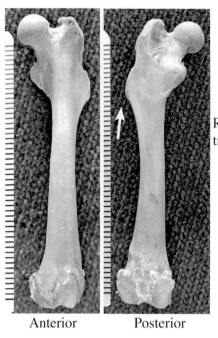
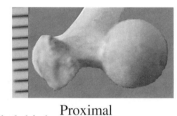
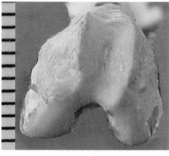

Anterior   Posterior

Rounded third trochanter

Proximal

Distal

Norway rat (*Rattus norvegicus*)

Proximal

Distal

The small round objects above the condyles are sesamoid bones and are not part of the femur.

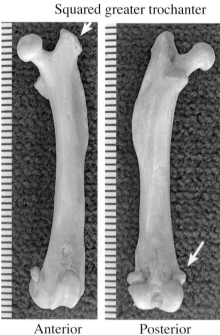

Squared greater trochanter

Anterior   Posterior

Femur

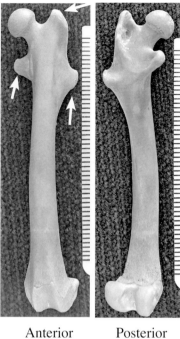

Proximal

Pronounced greater, lesser, and third trochanters

Anterior Posterior Distal

Squirrel (*Sciuridae sciurus niger*)

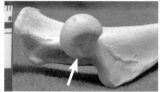

Semicircular notch Proximal

Pronounced greater, lesser, and third trochanters

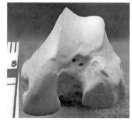
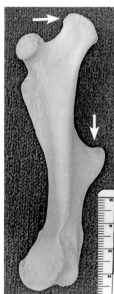
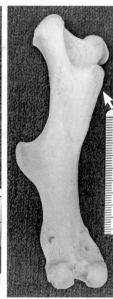

Distal Anterior Posterior

Armadillo (*Dasypus novemcinctus*)

Opossum (*Didelphis virginiana*)

Anterior

Posterior

Large lesser trochanter

Proximal

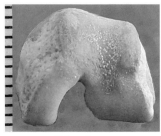
Distal

Very short femur and wide distally (leading to flipper)

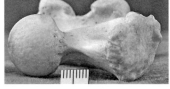
Proximal

Anterior articular surface takes up very little of distal aspect

Distal

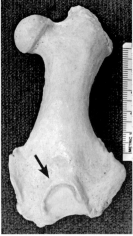
Anterior

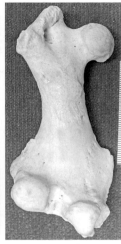
Posterior

Seal (*Phoca vitulina*)

Tibia

Features of the tibia 229
    Human (*Homo sapiens*) 231
    Moose (*Alces alces*) 232
    Elk (*Cervus elaphus*) 232
    Deer (*Odocoileus* sp.) 233
    Bison (*Bison bison*) 233
    Cow (*Bos taurus*) 234
    Antelope (*Antilocapra americana*) 234
    Mountain sheep (*Ovis canadensis*) 235
    Domestic sheep (*Ovis aries*) 235
    Domestic pig (*Sus scrofa*) 236
    Llama (*Lama glama*) 236
    Horse (*Equus*) 237
    Bear (*Ursus americanus*) 237
    Wolf (*Canis lupus*) 238
    Coyote (*Canis latrans*) 238
    Mountain lion (*Felis concolor*) 239
    Bobcat (*Lynx rufus*) 239
    Raccoon (*Procyon lotor*) 240
    Badger (*Taxidea taxus*) 240
    Skunk (*Mephitis mephitis*) 241
    River otter (*Lontra canadensis*) 241
    Rabbit (*Lepus* sp.) 242
    Beaver (*Castor canadensis*) 242
    Porcupine (*Erethizon dorsatum*) 243
    Marmot (*Marmota monax*) 243
    Prairie dog (*Cynomys gunnisoni*) 244
    Norway rat (*Rattus norvegicus*) 244
    Squirrel (*Sciuridae sciurus niger*) 245
    Armadillo (*Dasypus novemcinctus*) 245
    Opossum (*Didelphis virginiana*) 246
    Seal (*Phoca vitulina*) 246

Features of the Tibia

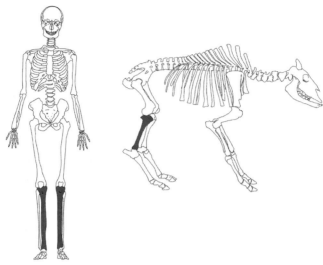

The tibia is the second-largest bone of the leg. The tibial tuberosity, which is anterior and near the proximal aspect of the bone, is the area of insertion for many of the muscles that flex the thigh at the hip and extend the leg. The medial malleolus is the part of the tibia in humans that forms the "bump" of the medial part of the ankle. It is more difficult to discern in some mammals. The distal tibia in nonhumans is often much more sculpted than in humans (though not always).

In general, the tibia and fibula are analogous to the radius and ulna. In humans, because the radius and ulna are separate and because the radius can rotate around the ulna to some degree, humans can pronate and supinate their forearms and hands. In humans the tibia and fibula are still two separate bones, but there is very little rotation of the lower leg and foot, so humans have lost the ability to pronate and supinate the lower limb. In many nonhumans (particularly large animals) the tibia and fibula are either fused or the fibula has become a vestigial bone or lost altogether, so no rotation is possible in the lower limb.

Distal aspect of human and nonhuman tibia. Note the more "sculpted" nature of the nonhuman tibia.

 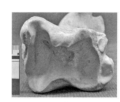

Human (*Homo sapiens*)   Moose (*Alces alces*)

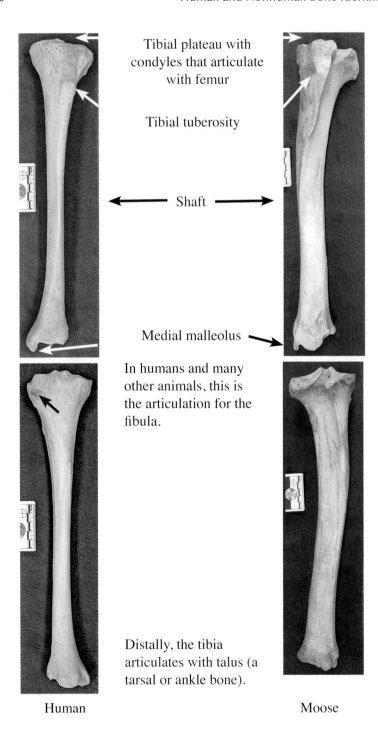

Tibia

Human (*Homo sapiens*)

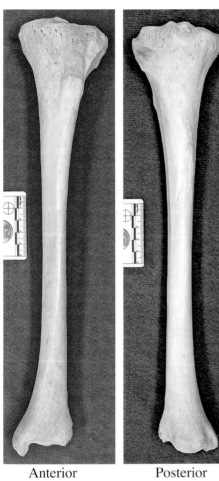

Anterior   Posterior

Anterior   Posterior

Proximal

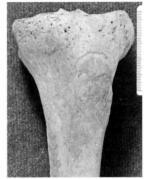

Proximal anterior

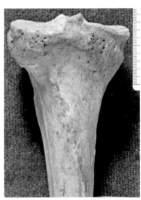

Proximal posterior

Distal

Moose (*Alces alces*)

Anterior

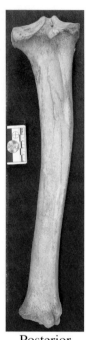
Posterior

Moose have no fibula.

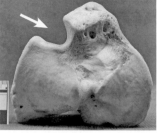
Notch
Proximal

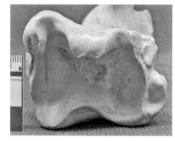
Distal

Elk (*Cervus elaphus*)

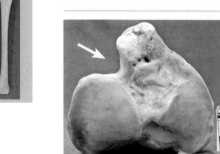
Notch
Proximal

Elk have no fibula.

Distal

Anterior

Posterior

Tibia

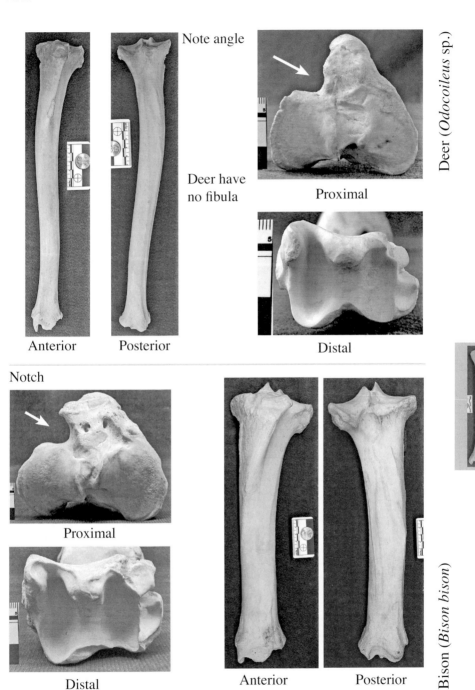

Bison and cow bones are short and massive and they have no fibula.

Cow (*Bos taurus*)

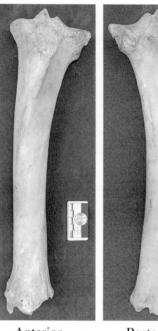

Anterior Posterior

Proximal

Distal

Bison and cow bones are short and massive and they have no fibula.

Antelope (*Antilocapra americana*)

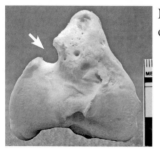
Proximal

Note shape of notch.

Note remnant fibula.

Distal

Antelope tibia is long and slender. They have only a remnant fibula.

Anterior Posterior

Tibia

Anterior Posterior

Note shape of notch.

Note small projection on posterior proximal tibia.

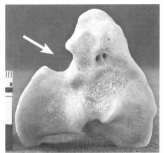

Proximal

Distal

Sheep have no fibula.

Mountain sheep (Ovis canadensis)

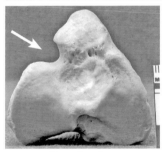

Proximal

Note shape of notch.

Note small projection on posterior proximal tibia.

Distal

Anterior Posterior

Sheep have no fibula.

Domestic sheep (Ovis aries)

Domestic pig (*Sus scrofa*)

Anterior — Posterior

Tibial tuberosity large and curved

Deep notch
Proximal

Distal

Llama (*Lama glama*)

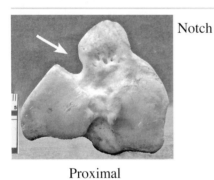
Notch
Proximal
Distal

Anterior — Posterior

Tibia

Anterior

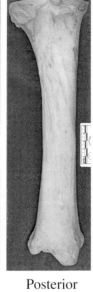

Posterior

Wide notch

Proximal

The horse fibula is a small splint.

Distal

Horse (Equus)

No notch

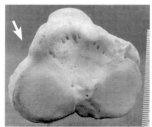

Proximal

Distal

Caution! Bear tibia can be confused with human!

Anterior   Posterior

Bear (Ursus americanus)

Wolf (*Canis lupus*)

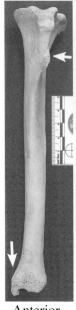

Anterior   Posterior

Notice shape of tibial tuberosity.

Prominent medial malleolus.

Notch

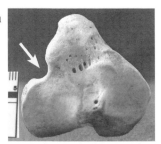

Proximal

Distal

Coyote (*Canis latrans*)

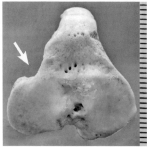

Proximal

Notch is different from wolf.

Distal

Coyote tibia is thinner relative to length than wolf.

Anterior   Posterior

Tibia

Anterior

Posterior

Prominent medial malleolus

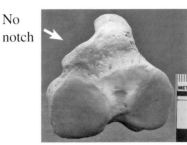
No notch

Proximal

Distal

Mountain lion (*Felis concolor*)

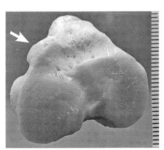
No notch

Proximal

Similar to mountain lion, but smaller

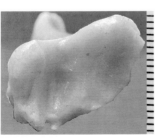
Distal

Anterior

Posterior

Bobcat (*Lynx rufus*)

Raccoon (*Procyon lotor*)

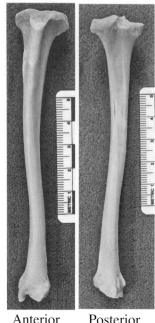

Anterior   Posterior

Proximal

Distal

Badger (*Taxidea taxus*)

Proximal

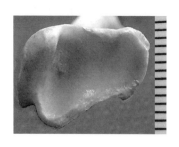

Distal

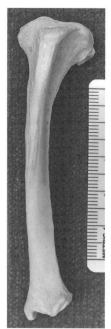 

Anterior   Posterior

Tibia

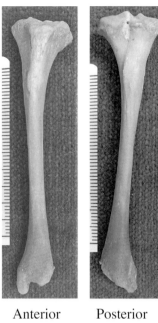

Anterior Posterior

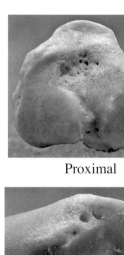

Proximal

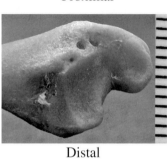

Distal

Skunk (*Mephitis mephitis*)

Proximal

Distal

Anterior Posterior

River otter (*Lontra canadensis*)

Rabbit (*Lepus* sp.)

Anterior   Posterior

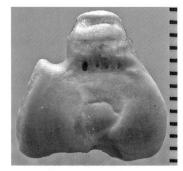

Proximal

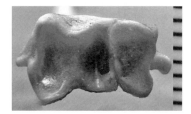

Distal

Beaver (*Castor canadensis*)

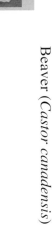

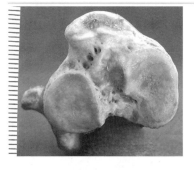

Distal

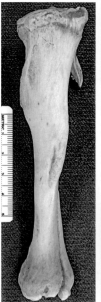
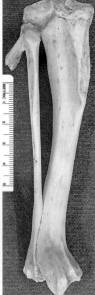

Anterior   Posterior

Tibia

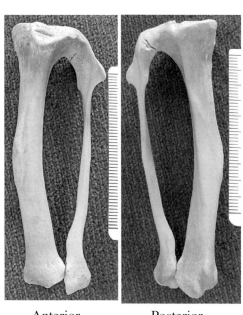

Anterior   Posterior

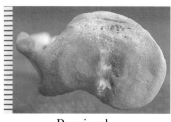

Proximal

Distal

Porcupine (*Erethizon dorsatum*)

Proximal

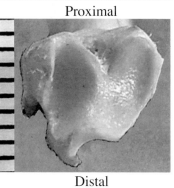

Distal

Anterior   Posterior

Marmot (*Marmota monax*)

Prairie dog (*Cynomys gunnisoni*)

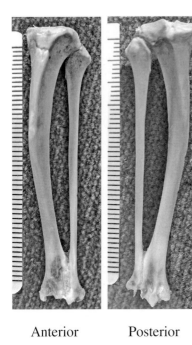

Anterior   Posterior

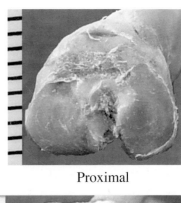

Proximal

Distal

Norway rat (*Rattus norvegicus*)

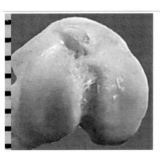

Proximal

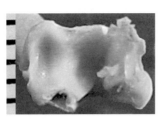

Distal

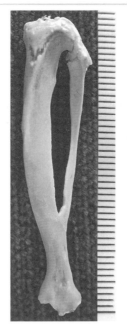 

Anterior   Posterior

Tibia

Anterior

Posterior

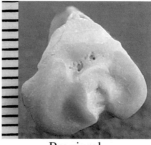
Proximal

Distal

Squirrel (*Sciuridae sciurus niger*)

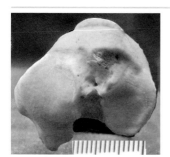
Proximal

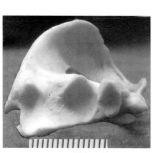
Distal

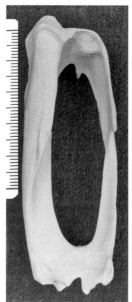
Anterior

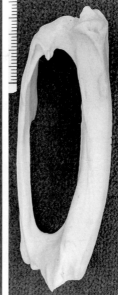
Posterior

Armadillo (*Dasypus novemcinctus*)

Opossum (*Didelphis virginiana*)

Anterior   Posterior

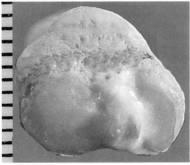

Proximal

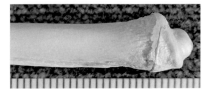

Distal

Seal (*Phoca vitulina*)

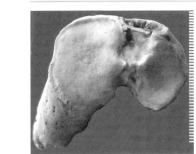

Proximal

Distal

 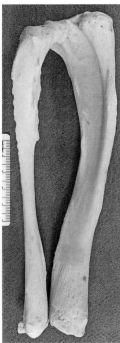

Anterior   Posterior

Fibula

Features of the fibula 249
    Human (*Homo sapiens*) 250
    Domestic pig (*Sus scrofa*) 251
    Bear (*Ursus americanus*) 251
    Wolf (*Canis lupus*) 252
    Coyote (*Canis latrans*) 252
    Mountain lion (*Felis concolor*) 253
    Bobcat (*Lynx rufus*) 253
    Raccoon (*Procyon lotor*) 254
    Badger (*Taxidea taxus*) 254
    Skunk (*Mephitis mephitus*) 255
    River otter (*Lontra canadensis*) 255
    Marmot (*Marmota monax*) 256
    Opossum (*Didelphis virginiana*) 256

Features of the Fibula

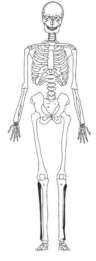

The bison (and many other large quadrupeds) have a vestigial or no fibula.

The fibula and tibia are analogous to the radius and ulna of the arm, although the pronation–supination function of the human arm is greatly reduced in the leg. The fibula has been lost, much reduced, or fused to the tibia in many quadrupeds. The lateral malleolus of the distal fibula forms the lateral projection of the ankle felt in humans. Distally, the fibula has a groove (the malleolar fossa) on the posterior surface. This is where a tendon crosses from the posterior leg to the plantar surface of the foot, and acts to, among other things, extend the foot. Proximally, the fibula articulates with the tibia.

Proximal articulates with tibia.

A. Human (*Homo sapiens*)
B. Pig (*Sus scrofa*)
C. Coyote (*Canis latrans*)

Distal articulates with talus (an ankle bone).

A

B

C

Human (*Homo sapiens*)

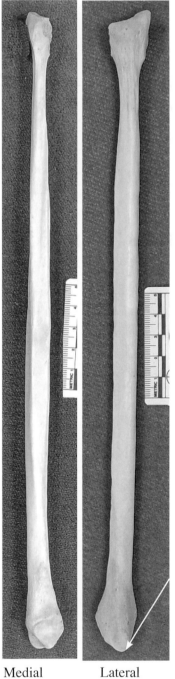

Medial   Lateral

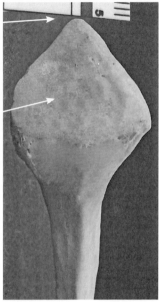

Styloid process

Articulates with tibia

Proximal medial view.

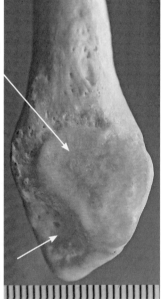

Articulates with talus

Lateral malleolus

Malleolar groove

Distal medial view.

Fibula

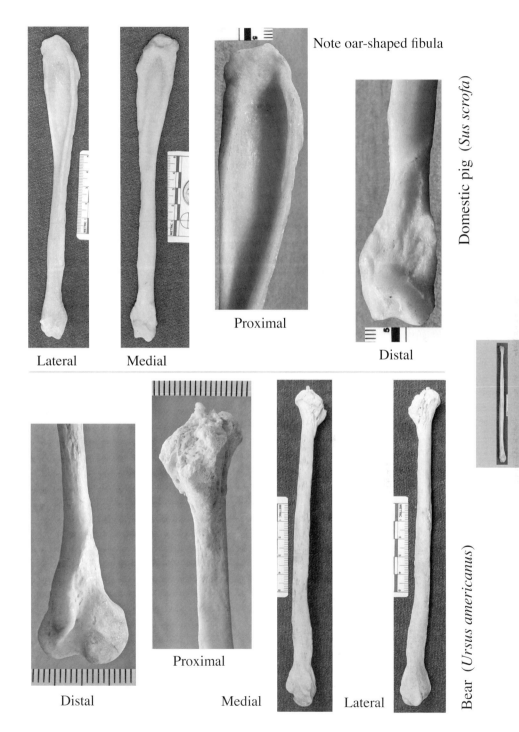

Note oar-shaped fibula

Domestic pig (*Sus scrofa*)

Lateral Medial Proximal Distal

Bear (*Ursus americanus*)

Distal Proximal Medial Lateral

Wolf (*Canis lupus*)

The coyote is more gracile than the wolf.

Coyote proximal medial view

Coyote distal medial view

Wolf Coyote lateral Coyote medial

Coyote (*Canis latrans*)

Fibula

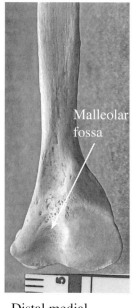
Distal medial

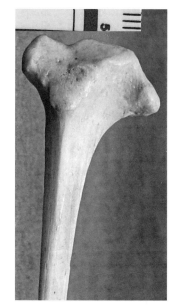
Proximal medial

Medial

Mountain lion (*Felis concolor*)

Lateral   Medial

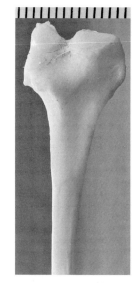
Proximal medial

Malleolar fossa — Distal medial

Bobcat (*Lynx rufus*)

254　　　　　　　　　　　　　　　　　　Human and Nonhuman Bone Identification

Raccoon (*Procyon lotor*)

Distal medial

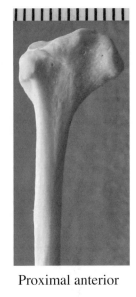

Proximal anterior

Malleolar fossa

Medial　　　Lateral

Badger (*Taxidea taxus*)

Lateral

Medial

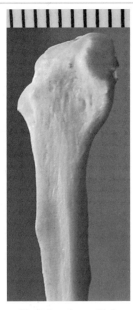

Proximal medial

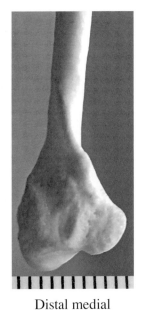

Distal medial

Fibula

Distal

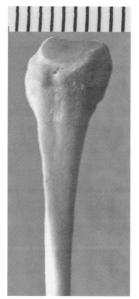
Proximal

Medial

Lateral

Skunk (*Mephitis mephitis*)

Lateral

Medial

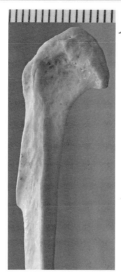
Proximal anterior

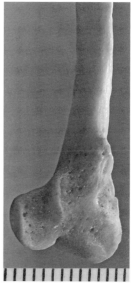
Distal medial

River otter (*Lontra canadensis*)

Marmot (*Marmota monax*)

Distal medial

Proximal anterior

Medial   Lateral

Opossum (*Didelphis virginiana*)

Lateral

Medial

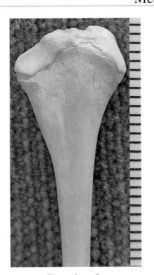
Proximal

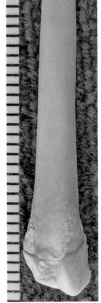
Distal

Metatarsals and Hindlimbs

Features of the metatarsals and hindlimbs 259
    Human (*Homo sapiens*) 262
    Moose (*Alces alces*) 263
    Elk (*Cervus elaphus*) 263
    Deer (*Odocoileus* sp.) 264
    Bison (*Bison bison*) 264
    Cow (*Bos taurus*) 265
    Antelope (*Antilocapra americana*) 265
    Mountain sheep (*Ovis canadensis*) 266
    Domestic sheep (*Ovis aries*) 266
    Domestic pig (*Sus scrofa*) 267
    Llama (*Lama glama*) 267
    Horse (*Equus*) 268
    Bear (*Ursus americanus*) 268
    Coyote (*Canis latrans*) 269
    Bobcat (*Lynx rufus*) 269
    Norway rat (*Rattus norvegicus*) 270
    Squirrel (*Sciuridae sciurus niger*) 270
    Armadillo (*Dasypus novemcinctus*) 271
    Vampire bat (*Vampyressa nymphaea*) 271

Features of the Metatarsals and Hindlimbs

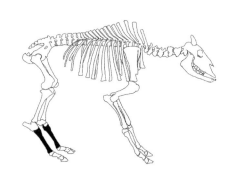

Note: Some of the basic information presented in this section is either repeated or analogous to that presented in the Metacarpal section.

The metatarsals are the bones that make up the main part of the foot in the human (minus the toes), and part of the leg (in large animals) or paw (in smaller animals). The number and morphology of metatarsals in humans and nonhumans have everything to do with how the feet, hindpaws, and hindlimbs are used (primarily modes of locomotion).

An array of five digits (for the hand or paw) is the most primitive condition. Many large animals have a reduced number to two (for example in the elk, cow, sheep, etc.) and to one (as in the horse).

The hindlimbs in humans and most other animals are responsible for locomotion (the bat uses its forelimbs for locomotion). Note on the following page that even the alignment of the metatarsals and phalanges is different in different animals. In many animals with paws the metatarsals and phalanges are very tightly grouped, and even in humans the bones in the foot are more tightly grouped than in the hand (the first metatarsal does not diverge from the rest of the metatarsals the way the first metacarpal diverges away from the rest of the metacarpals in the hand). This information can be important when trying to determine human from nonhuman, particularly in tissued specimens or in an x-ray.

Caution! It is possible (if not common!) to mistake a bear paw for a human hand or foot. Please see page 165 for details.

260 Human and Nonhuman Bone Identification

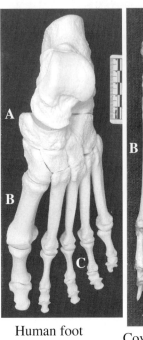

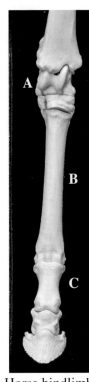

Human foot   Coyote paw

Moose hindlimb   Horse hindlimb

A. Tarsals
B. Metatarsals
C. Phalanges

Rat hindlimb   Bat foot   Armadillo hand

Note that there are three rows of phalanges: proximal, middle, and distal. Distal phalanges are usually claws in small mammals and hooves in large mammals.

Matatarsals and Hindlimbs

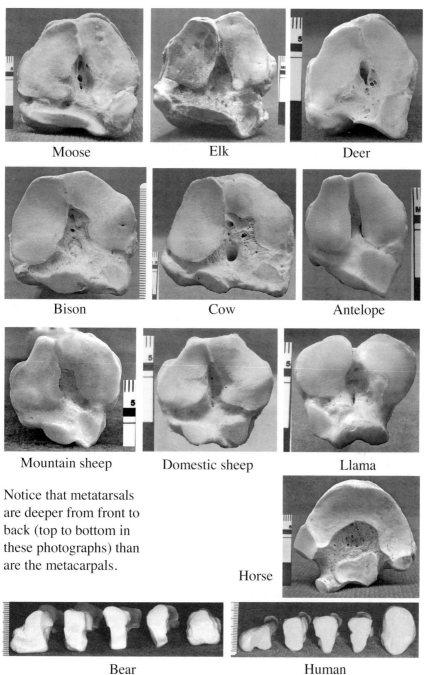

Moose

Elk

Deer

Bison

Cow

Antelope

Mountain sheep

Domestic sheep

Llama

Notice that metatarsals are deeper from front to back (top to bottom in these photographs) than are the metacarpals.

Horse

Bear

Human

Proximal metatarsals

Human (*Homo sapiens*)

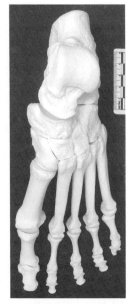

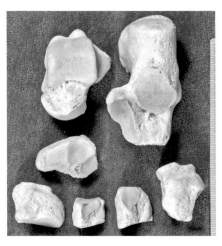

Tarsals

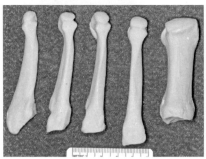

Metatarsals superior

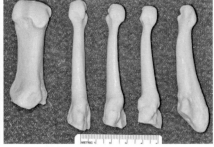

Metatarsals plantar

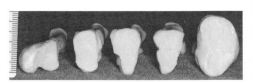

Metatarsals proximal view

Foot phalanges

Matatarsals and Hindlimbs

Anterior — Posterior

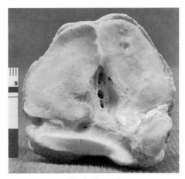

Proximal

Distal

Moose (*Alces alces*)

Proximal

Distal

Anterior — Posterior

Elk (*Cervus elaphus*)

Deer (*Odocoileus* sp.)

Anterior

Posterior

← No posterior groove

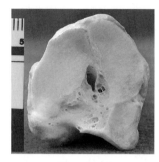
Proximal

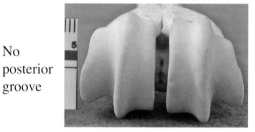
Distal

Bison (*Bison bison*)

Proximal

Distal

Anterior

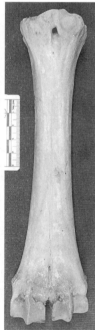
Posterior

Matatarsals and Hindlimbs

Anterior   Posterior

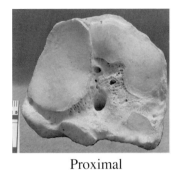

Proximal

Groove extends beyond foramen.

Distal

Cow (Bos taurus)

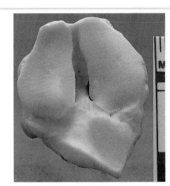

Proximal

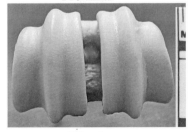

Distal

Anterior   Posterior

Antelope (Antilocapra americana)

Mountain sheep (*Ovis canadensis*)

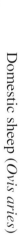

Proximal

← No well-defined groove

Anterior   Posterior

Domestic sheep (*Ovis aries*)

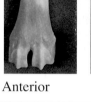

Proximal

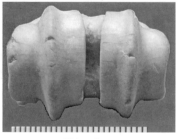

Distal

Anterior   Posterior

Matatarsals and Hindlimbs

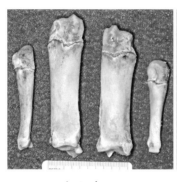

Anterior

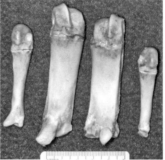

Posterior

Ham hock elements are frequently confused with human foot bones. Remember to look for the ridge in the middle of the distal articular surface of the metatarsals.

Superior phalanges

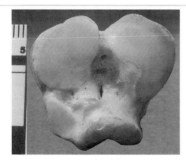

Proximal

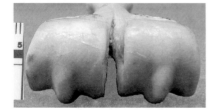

Phalanges

Complete separation of condyles

Anterior

Posterior

Domestic pig (*Sus scrofa*)

Llama (*Lama glama*)

Horse (*Equus*)

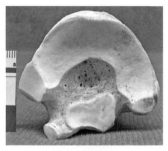
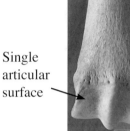

Anterior Posterior

Proximal

Single articular surface

Distal

Bear (*Ursus americanus*)

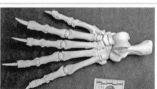
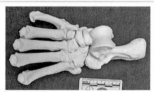

Bear hindpaw with claws   Bear hindpaw with no claws

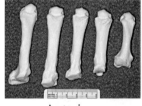
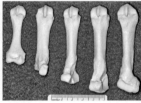

Anterior Posterior Tarsals

Proximal

Phalanges

Matatarsals and Hindlimbs 269

Anterior Posterior

Coyote (Canis latrans)

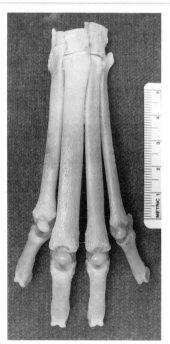
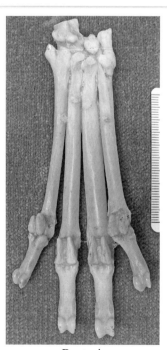

Anterior Posterior

Bobcat (Lynx rufus)

Norway rat (*Rattus norvegicus*)

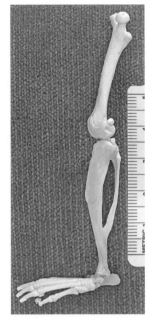

Medial view

Squirrel (*Sciuridae sciurus niger*)

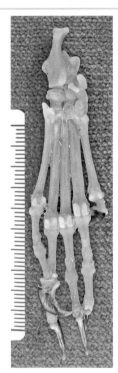

Matatarsals and Hindlimbs

Anterior

Posterior

Armadillo (*Dasypus novemcinctus*)

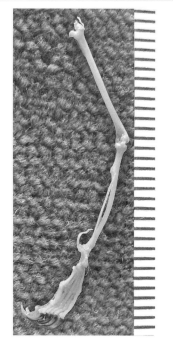

Vampire bat (*Vampyressa nymphaea*)

Index

A

Accipiter cooperii (Cooper's hawk), 35
acetabulum, 209
achondroplasty, 28
acromion process, 90
alveoli, 30
antelope (*Antilocapra americana*)
    cranium, 53
    femur, 214
    humerus, 114
    mandible, 73
    metacarpals and forelimbs, 174, 178
    metatarsals and hindlimbs, 261, 265
    pelvic girdle, 189, 194
    radius, 132, 135
    scapula, 93
    tibia, 234
    ulna, 156
anterior, defined, 15, 16
Antilocapra americana. See Antelope
appendicular, defined, 16
Aquila chrysaetos (golden eagle), 34
armadillo (*Dasypus novemcinctus*)
    cranium, 64
    femur, 225
    humerus, 125
    mandible, 84
    metatarsals and hindlimbs, 260, 271
    pelvic girdle, 205
    radius, 132, 146
    scapula, 104
    tibia, 245
    ulna, 167
articular surface, 12, 13, 24, 25
artiodactyls, 29
ascending ramus, 69, 70
Athene cunicularia (burrowing owl), 35
axial, defined, 16

B

badger (*Taxidea taxus*)
    cranium, 59
    femur, 220
    fibula, 254
    humerus, 120
    mandible, 79
    metacarpals and forelimbs, 183
    pelvic girdle, 200
    radius, 132, 141
    scapula, 99
    tibia, 240
    ulna, 162
bald eagle (*Haliaeetus leucocephalus*), 34
bat, 252. *See also* Vampire bat
bear (*Ursus americanus*)
    cranium, 56
    femur, 217
    fibula, 251
    humerus, 117
    mandible, 76
    metacarpals and forelimbs, 174, 181
    metatarsals and hindlimbs, 259, 261, 268
    paw, 173
    pelvic girdle, 197
    radius, 132, 138
    scapula, 96
    tibia, 237
    ulna, 159

beaver (*Castor canadensis*)
  cranium, 61
  femur, 210, 222
  humerus, 110, 122
  mandible, 81
  pelvic girdle, 202
  radius, 132, 143
  scapula, 101
  tibia, 242
  ulna, 164
beetles, dermestid, 41–42
bipeds, overview of, 17–18
birds, 8, 31–40. *See also Specific birds*
bison (*Bison bison*)
  cranium, 52
  femur, 213
  fibula, 249
  humerus, 113
  mandible, 72
  metacarpals and forelimbs, 174, 187
  metatarsals and hindlimbs, 261, 264
  pelvic girdle, 190, 193
  radius, 132, 134
  scapula, 92
  skeleton, 7
  spinous process, 20
  tibia, 233
  ulna, 155
bobcat (*Lynx rufus*)
  cranium, 58
  femur, 219
  fibula, 253
  humerus, 119
  mandible, 78
  metatarsals and hindlimbs, 269
  pelvic girdle, 199
  radius, 132, 140
  scapula, 98
  tibia, 239
  ulna, 161
bone. *See also Specific bones*
  bird, 31

  cleaning and storage of, 41–43
  color of, 9–10
  dermestid beetles and, 41–42
  growth and development of, 25–29
  morphology of, 11–14
  overview of, 10–11
  overview of skeletal, 18–25
  preservation of, 43
  quadruped skeletal morphology and, 18
  terminology of, 15–17
Bos taurus. *See* Cow
brachiators, 18
Bubo virginianus (great horned owl), 34
buccal, defined, 16
burrowing owl (*Athene cunicularia*), 35
Buteo jamaicensis (red-tailed hawk), 35
Buteo lineatus (red-shouldered hawk), 38, 39, 40

C

cancellous bone, 14
Canis domesticus. *See* Dog
Canis latrans. *See* Coyote
Canis lupus. *See* Wolf
carpals. *See* Metacarpals and forelimbs
cartilage, 12, 26
cartilagic model, 26–29
Castor canadensis. *See* Beaver
cat, domestic (*Felis domesticus*), 58
Cathartes aura septentrionalis (turkey vulture), 32–33, 35
caudal, defined, 15, 16
cavities, defined, 17
cervical vertebrae, 18, 19, 37
Cervus elaphus. *See* Elk
chest. *See* Thorax
chicken, 28, 40
clavicle, 20–21
cleaning of bone, 41
coccygeal vertebrae, 18

collagen, 10
color of bone, 9–10
condyles, 17, 42, 69, 209
Cooper's hawk (*Accipiter cooperii*), 35
coracoid bone, 36
coronal plane, defined, 16
cortical bone, 14
cow (*Bos taurus*)
   cranium, 53
   femur, 31, 214
   humerus, 114
   mandible, 73
   metacarpals and forelimbs, 174, 178
   metatarsals and hindlimbs, 261, 265
   pelvic girdle, 194
   radius, 132, 135
   ribs, 20
   scapula, 93
   tibia, 234
   ulna, 151, 156
coyote (*Canis latrans*)
   cranium, 57
   femur, 218
   fibula, 249, 252
   humerus, 118
   mandible, 77
   metatarsals and hindlimbs, 260, 269
   pelvic girdle, 198
   radius, 132, 139
   scapula, 97
   tibia, 238
   ulna, 160
cranial, defined, 15, 16
cranium, 21–22, 41
   antelope, 53
   armadillo, 64
   badger, 59
   bear, 56
   beaver, 61
   bison, 52
   bobcat, 58
   cat, domestic, 58
   cow, 53
   coyote, 57
   deer, 52

   dog, 57
   elk, 51
   horse, 56
   human, 50
   human vs. moose, 14
   llama, 55
   marmot, 62
   moose, 51
   Norway rat, 63
   opossum, 65
   pig, 55
   porcupine, 62
   prairie dog, 63
   rabbit, 61
   raccoon, 59
   river otter, 60
   seal, 65
   sheep (mountain), 54
   sheep, domestic, 54
   skunk, 60
   squirrel, 64
   turkey vulture, 32–33
   wolf, 57
crests, defined, 17
crypts, 30
Cynomys gunnisoni. See Prairie dog

D

Dasypus novemcinctus. See Armadillo
deciduous teeth, 30
deer (*Odocoileus* sp.)
   cranium, 52
   femur, 213
   humerus, 113
   mandible, 72
   metacarpals and forelimbs, 172, 177
   metatarsals and hindlimbs, 261, 264
   osteons, 29
   pelvic girdle, 193
   radius, 132, 134
   scapula, 92
   tibia, 233
   ulna, 11, 155
deltoid tuberosity, 109, 110

dentition, 30, 69
dermestid beetles, 41–42
development, overview of, 25–29
dew claws, 25
diaphyses, 17, 26
Didelphis virginiana. *See* Opossum
distal, defined, 15, 16
dog (*Canis domesticus*), 6, 57
dorsal, defined, 15, 16
dwarfism, 28

E

eagles, 34
eburnation, 13
elk (*Cervus elaphus*)
   cranium, 51
   femur, 212
   humerus, 112
   mandible, 71
   metacarpals and forelimbs, 174, 176
   metatarsals and hindlimbs, 261, 263
   pelvic girdle, 192
   radius, 132, 133
   scapula, 91
   tibia, 232
   ulna, 154
epiphyses, 17, 26, 27
Equus. *See* Horse
Erethizon dorsatum. *See* Porcupine
external, defined, 16

F

Felis domesticus. *See* Cat, domestic
femur, 209
   antelope, 214
   armadillo, 225
   badger, 220
   bear, 217
   beaver, 220, 222
   bison, 213
   bobcat, 219
   cow, 31, 214
   coyote, 218
   deer, 213
   development of human, 27, 28
   elk, 212
   horse, 217
   human, 11, 209, 210, 211
   llama, 216
   marmot, 223
   moose, 11, 209, 210, 212
   mountain lion, 219
   Norway rat, 224
   opossum, 226
   pig, 216
   porcupine, 223
   prairie dog, 2224
   rabbit, 222
   raccoon, 220
   river otter, 221
   seal, 226
   sheep, domestic, 225
   sheep, mountain, 225
   skunk, 221
   squirrel, 225
   turkey, 31, 40
   wolf, 218
fibula, 24, 249
   badger, 254
   bear, 251
   bison, 249
   bobcat, 253
   coyote, 249, 252
   human, 249, 250
   marmot, 256
   mountain lion, 253
   opossum, 256
   pig, 241, 251
   raccoon, 254
   river otter, 255
   skunk, 255
   wolf, 252
fontanelles, defined, 17
foot, 24–25. *See also* Metatarsals and hindlimbs
foramen, 27

foramen magnum, 21, 22, 41, 42
foramina, defined, 17
forelimbs, 23–25. *See also*
Metacarpals and forelimbs
fossae, defined, 17
fragmentary osteons, 29
furcula, turkey, 37

G

giraffe, 19
glenoid fossa, 89, 90, 1110
golden eagle (*Aquila chrysaetos*), 34
gravity, center of, 22–23
great horned owl (*Bubo virginianus*), 34
growth and development, 25–29
growth plates, 26, 27

H

Haliaeetus leucocephalus (bald eagle), 34
hand, 24–25. *See also* Metacarpals and forelimbs
hawks, 35, 38, 39, 40
hindlimbs, 23–25. *See also*
Metatarsals and hindlimbs
hip joint. *See* Os coxae
horse (*Equus*)
    cranium, 56
    femur, 217
    hoof, 25
    humerus, 117
    mandible, 76
    metacarpals and forelimbs, 174, 181
    metatarsals and hindlimbs, 260, 261, 268
    pelvic girdle, 197
    radius, 132, 138
    scapula, 96
    tibia, 237
    ulna, 159
human (*Homo sapiens*)
    articular surface, 25

cervical vertebrae, 19
clavicle, 20
cranium, 14, 42
dentition, 30
femur, 11, 27, 28, 209, 210, 211
fibula, 249, 250
foramen magnum, 22
hand, 25, 172, 173
humerus, 109, 110, 111
mandible, 69, 70
mastoid process, 22
metacarpals and forelimbs, 174, 175
metatarsals and hindlimbs, 259, 260, 261, 262
osteons, 29
pelvic girdle, 23, 189, 190, 191
radius, 24, 130, 131, 132
ribs, 20
scapula, 21, 89, 90
skeleton, 5
spinous process, 20
tibia, 229, 230, 231
ulna, 11, 24, 151, 152, 153
humeral head, 109, 110
humerus, 109
    antelope, 114
    armadillo, 125
    badger, 120
    bear, 118
    beaver, 110, 122
    bison, 113
    bobcat, 119
    cow, 114
    coyote, 118
    deer, 113
    elk, 112
    horse, 117
    human, 109, 110, 111
    llama, 116
    marmot, 123
    moose, 110, 112
    mountain lion, 119
    Norway rat, 124
    opossum, 126

pig, 116
porcupine, 123
prairie dog, 124
rabbit, 122
raccoon, 120
river otter, 121
seal, 126
sheep, domestic, 115
sheep, mountain, 115
skunk, 121
squirrel, 125
turkey, 39
wolf, 118
hydroxyapatite, 10

I

ilium, 189. *See also* Pelvic girdle
inferior, defined, 16
innominates. *See* Os coxae
internal, defined, 16
ischium, 189. *See also* Pelvic girdle

J

jaw, lower. *See* Mandible

L

labial, defined, 16
lacrimal pits, 51
lateral, defined, 15, 16
Lepus sp. *See* Rabbit
lever arms, 23–24
limbs, overview of, 23–25
lingual, defined, 16
llama (*Llama glama*)
    cranium, 55
    femur, 216
    humerus, 116
    mandible, 75
    metacarpals and forelimbs, 174, 180
    metatarsals and hindlimbs, 261, 267
    pelvic girdle, 196

radius, 132, 137
scapula, 95
tibia, 236
ulna, 158
locomotion, skeletal morphology and, 18
long bones, 14
longitudinal, defined, 16
Lontra canadensis. *See* River otter
lumbar vertebrae, 18
Lynx rufus. *See* Bobcat

M

malleolar fossa, 249
malleolus, lateral, 249
malleolus, medial, 229, 230
mandible, 30, 69
    antelope, 73
    armadillo, 74
    badger, 79
    bear, 75
    beaver, 81
    bison, 72
    bobcat, 78
    cow, 73
    coyote, 77
    deer, 72
    elk, 71
    horse, 75
    human, 69, 70
    llama, 75
    marmot, 82
    moose, 69, 71
    mountain lion, 78
    Norway rat, 83
    opossum, 85
    pig, 75
    porcupine, 82
    prairie dog, 83
    rabbit, 81
    raccoon, 79
    river otter, 80
    seal, 85
    sheep, domestic, 74

sheep, mountain, 74
skunk, 80
squirrel, 84
wolf, 77
marmot (*Marmota monax*)
    cranium, 62
    femur, 223
    fibula, 256
    humerus, 123
    mandible, 82
    metacarpals and forelimbs, 185
    pelvic girdle, 203
    radius, 132, 144
    scapula, 102
    tibia, 243
    ulna, 165
masseter muscle, 69
mastoid process, 21–22, 41, 42
meatuses, defined, 17
medial, defined, 15, 16
medial malleolus, 229, 230
Mephitis mephitis. *See* Skunk
mesial, defined, 16
metacarpals and forelimbs, 171–174
    antelope, 174, 178
    badger, 183
    bear, 173, 174, 181
    bison, 174, 177
    cow, 174, 178
    deer, 174, 177
    elk, 174, 176
    horse, 174, 181
    human, 172, 173, 174, 175
    llama, 174, 180
    marmot, 185
    moose, 172, 174, 176
    mountain lion, 182
    Norway rat, 185
    pig, 180
    rabbit, 184
    raccoon, 183
    rat, 172
    seal, 172, 186
    sheep, domestic, 174, 179
    sheep, mountain, 174, 179
    skunk, 172, 184
    vampire bat, 186
    wolf, 172, 182
metatarsals and hindlimbs, 259–261
    antelope, 261, 265
    armadillo, 260, 271
    bat, 260
    bear, 259, 261, 268
    bison, 261, 264
    bobcat, 269
    cow, 261, 265
    coyote, 260, 269
    deer, 261, 264
    elk, 261, 263
    horse, 260, 261, 268
    human, 259, 260, 261, 262
    llama, 261, 267
    moose, 260, 261, 263
    Norway rat, 270
    pig, 267
    rat, 260
    sheep, domestic, 261, 266
    sheep, mountain, 261, 266
    squirrel, 270
    vampire bat, 271
midsagittal plane, defined, 15, 16
moose (*Alces alces*)
    articular surface, 25
    cranium, 14, 43
    femur, 11, 209, 210, 212
    foramen magnum, 22
    forelimb, 172
    humerus, 110, 112
    mandible, 69, 71
    metacarpals and forelimbs, 174, 176
    metatarsals and hindlimbs, 260, 261, 263
    pelvic girdle, 192
    radius, 130, 131, 132
    scapula, 21, 91
    tibia, 229, 230, 232
    ulna, 152, 154
morphology, overview of, 11–14
mountain lion (*Felis concolor*)

femur, 219
fibula, 253
humerus, 119
mandible, 78
metacarpals and forelimbs, 182
pelvic girdle, 199
radius, 132, 140
scapula, 98
tibia, 239
ulna, 161
vertebral column, 19
muscle insertion areas, 12–13

N

Norway rat (*Rattus norvegicus*)
    cranium, 63
    femur, 224
    humerus, 124
    mandible, 83
    metacarpals and forelimbs, 185
    metatarsals and hindlimbs, 270
    pelvic girdle, 204
    radius, 132, 145
    scapula, 103
    tibia, 244
    ulna, 166
nuchal line (crest), 41
nutrient foramen, 27

O

occipital condyle, 41, 42
occlusal, defined, 16
Odocoileus sp. *See* Deer
olecranon process, 24, 152, 153
opossum (*Didelphis virginiana*)
    cranium, 65
    femur, 226
    fibula, 256
    humerus, 126
    mandible, 85
    pelvic girdle, 206
    radius, 132, 147
    scapula, 105

tibia, 246
ulna, 168
os coxae, 22, 38, 189, 190, 191, 209.
    See also Pelvic girdle
osteoblasts, 27, 29
osteocytes, 29
osteogenesis, 26–27
osteons, 29
otter. *See* River otter
Ovis aries. *See* Sheep, domestic
Ovis canadensis. *See* Sheep, mountain
owls, 34, 35

P

paws. *See* Metacarpals and forelimbs;
    Metatarsals and hindlimbs
pelvic girdle, 22–23, 189
    antelope, 189, 194
    armadillo, 205
    badger, 200
    bear, 197
    beaver, 202
    bison, 190, 193
    bobcat, 199
    cow, 194
    coyote, 198
    deer, 193
    elk, 192
    horse, 197
    human, 189, 190, 191
    llama, 196
    marmot, 203
    moose, 191
    mountain lion, 199
    Norway rat, 204
    opossum, 206
    pig, 196
    porcupine, 203
    prairie dog, 204
    rabbit, 202
    raccoon, 200
    river otter, 201
    seal, 206
    sheep, domestic, 195

sheep, mountain, 195
skunk, 201
squirrel, 205
wolf, 198
pelvis, 22–23
phalanges. *See* Metacarpals and forelimbs; Metatarsals and hindlimbs
pig (*Sus scrofa*)
  cranium, 55
  femur, 216
  fibula, 249, 251
  humerus, 116
  mandible, 75
  metacarpals and forelimbs, 180
  metatarsals and hindlimbs, 267
  pelvic girdle, 23, 196
  radius, 132, 137
  scapula, 85
  tibia, 236
  ulna, 158
planes of body, overview of, 15
plexiform bone, 29
pneumatic bones, 31
polishing, 13
porcupine (*Erethizon dorsatum*)
  cranium, 62
  femur, 223
  humerus, 123
  mandible, 82
  pelvic girdle, 203
  radius, 132, 144
  scapula, 102
  tibia, 243
  ulna, 165
posterior, defined, 15, 16
prairie dog (*Cynomys gunnisoni*)
  cranium, 63
  femur, 224
  humerus, 124
  mandible, 83
  pelvic girdle, 204
  radius, 132, 145
  scapula, 103
  tibia, 244
  ulna, 166
preservation, 51
processes, defined, 17
Procyon lotor. *See* Raccoon
pronation, defined, 16
proximal, defined, 15, 16
pubis, 189. *See also* Pelvic girdle
quadrupeds, overview of, 17–18

R

rabbit (*Lepus* sp.)
  cranium, 61
  femur, 222
  humerus, 122
  mandible, 81
  metacarpals and forelimbs, 184
  pelvic girdle, 1202
  radius, 132, 143
  scapula, 101
  tibia, 242
  ulna, 164
raccoon (*Procyon lotor*)
  cranium, 59
  femur, 220
  fibula, 254
  humerus, 120
  mandible, 79
  metacarpals and forelimbs, 183
  pelvic girdle, 200
  radius, 132, 141
  scapula, 99
  tibia, 240
  ulna, 162
radial notch, 152
radius, 24, 129
  antelope, 132, 135
  armadillo, 132, 146
  badger, 132, 141
  bear, 132, 138
  beaver, 132, 143
  bison, 132, 134
  bobcat, 132, 140
  cow, 132, 135
  coyote, 132, 139
  deer, 132, 134

elk, 132, 133
horse, 132, 138
human, 130, 131, 132
llama, 132, 137
marmot, 132, 144
moose, 130, 132, 133
mountain lion, 132, 140
Norway rat, 132, 145
opossum, 132, 147
pig, 132, 137
porcupine, 132, 144
prairie dog, 132, 145
rabbit, 132, 143
raccoon, 132, 141
river otter, 132, 142
seal, 132, 147
sheep, domestic, 132, 136
sheep, mountain, 132, 136
skunk, 32, 142
squirrel, 132, 146
turkey, 39
wolf, 132, 139
raptors, skulls of, 34–35
rat, 172, 260. *See also* Norway rat
red-shouldered hawk (*Buteo lineatus*), 38, 39, 40
red-tailed hawk (*Buteo jamaicensis*), 35
remodeling, 29
ribs, 20, 37. *See also* Thorax
river otter (*Lontra canadensis*)
   cranium, 60
   femur, 221
   fibula, 255
   humerus, 121
   mandible, 80
   pelvic girdle, 201
   radius, 132, 142
   scapula, 100
   tibia, 241
   ulna, 163

S

sacral vertebrae, 18
sacrum, 22, 189, 190, 191. *See also* Pelvic girdle
sagittal section, defined, 16
scapula, 21, 89
   antelope, 93
   armadillo, 104
   badger, 99
   bear, 96
   beaver, 101
   bison, 92
   bobcat, 98
   cow, 93
   coyote, 97
   deer, 92
   elk, 91
   horse, 96
   human, 89, 90
   llama, 95
   marmot, 102
   moose, 91
   mountain lion, 98
   Norway rat, 103
   opossum, 105
   pig, 95
   porcupine, 102
   prairie dog, 103
   rabbit, 101
   raccoon, 99
   river otter, 100
   seal, 105
   sheep, domestic, 94
   sheep, mountain, 94
   skunk, 100
   squirrel, 104
   turkey, 36
   wolf, 97
Sciuridae sciurus niger. *See* Squirrel
seal (*Phoca vitulina*)
   cranium, 65
   femur, 226
   hand, 172
   humerus, 126
   mandible, 85
   metacarpals and forelimbs, 186
   pelvic girdle, 206

Index

radius, 132, 147
scapula, 105
tibia, 246
ulna, 168
semilunar notch, 152
sheep, domestic (*Ovis aries*)
    cranium, 54
    femur, 215
    humerus, 115
    mandible, 74
    metacarpals and forelimbs, 174, 179
    metatarsals and hindlimbs, 261, 266
    pelvic girdle, 195
    radius, 132, 136
    scapula, 94
    tibia, 235
    ulna, 157
sheep, mountain (*Ovis canadensis*)
    cranium, 54
    femur, 215
    humerus, 115
    mandible, 74
    metacarpals and forelimbs, 174, 179
    metatarsals and hindlimbs, 261, 266
    pelvic girdle, 195
    radius, 132, 132
    scapula, 94
    tibia, 235
    ulna, 157
shoulder blade. *See* Scapula
sinuses, defined, 17
skeleton
    bird, 8, 31–40
    bison, 7
    dog, 6
    human, 5
    quadruped, 17–18
skulls, bird, 34–35
skunk (*Mephitis mephitis*)
    cranium, 60
    femur, 221
    fibula, 255
    humerus, 121
    mandible, 80
    metacarpals and forelimbs, 184

    paw, 172
    pelvic girdle, 201
    radius, 132, 142
    scapula, 100
    tibia, 241
    ulna, 163
smell, sense of, 21
spinal cord, 41, 42
spinous process, 19–20
spongy bone, 14
squirrel (*Sciuridae sciurus niger*)
    cranium, 64
    femur, 225
    humerus, 125
    mandible, 84
    metatarsals and hindlimbs, 270
    pelvic girdle, 205
    radius, 132, 146
    scapula, 104
    tibia, 245
    ulna, 167
sternocleidomastoid muscle, 22
sternum, turkey, 36
storage of bone, 42–43
styloid process, 152, 250
superficial, defined, 16
superior, defined, 16
supination, defined, 16
supratrochlear foramen, 111
Sus scrofa. See Pig
sutures, defined, 17
synovial fluid, 12

T

taphonomy, defined, 10
tarsals. *See* Metatarsals and hindlimbs
tarsometatarsus, 40
Taxidea taxus. See Badger
teeth, 30, 69
thoracic vertebrae, 18, 37
thorax, 20–21
tibia, 24, 229
    antelope, 234
    armadillo, 245

badger, 240
bear, 237
beaver, 242
bison, 233
bobcat, 239
cow, 234
coyote, 238
deer, 233
elk, 232
horse, 237
human, 229, 230, 231
llama, 236
marmot, 243
moose, 229, 230, 232
mountain lion, 239
Norway rat, 244
opossum, 246
pig, 236
porcupine, 243
prairie dog, 244
rabbit, 242
raccoon, 240
river otter, 241
seal, 246
sheep, domestic, 235
sheep, mountain, 235
skunk, 241
squirrel, 245
wolf, 238
tibial tuberosity, 229, 230
tibiotarsus, turkey, 40
tori, defined, 17
trabecular bone, 14
transverse, defined, 16
tubercles, defined, 17
tuberosities, defined, 17
turkey (*Meleagris* sp.), 31, 36–40
turkey vulture (*Cathartes aura septentrionalis*), 32–33, 35

U

ulna, 24, 150
  antelope, 156
  armadillo, 167
  badger, 162
  bear, 159
  beaver, 164
  bison, 155
  bobcat, 161
  cow, 151, 156
  coyote, 160
  deer, 155
  elk, 154
  horse, 159
  human, 151, 152, 153
  human vs. deer, 11
  llama, 158
  marmot, 165
  moose, 152, 154
  mountain lion, 161
  Norway rat, 166
  opossum, 168
  pig, 158
  porcupine, 165
  prairie dog, 166
  rabbit, 164
  raccoon, 162
  river otter, 163
  seal, 168
  sheep, domestic, 157
  sheep, mountain, 157
  skunk, 163
  squirrel, 167
  turkey, 39
  wolf, 151, 160
ulnar notch, 129, 130
Ursus americanus. See Bear

V

vampire bat (*Vampyressa nymphaea*), 186, 271
ventral, defined, 15, 16
vertebral column, 18–20, 37–38
vertex, defined, 16
vulture, turkey (*Cathartes aura septentrionalis*), 32–33, 35

water cleaning methods for bone, 42
wishbone, 37
wolf (*Canis lupus*)
   cranium, 57
   femur, 218
   fibula, 252
   humerus, 118
   mandible, 77
   mastoid process, 22
   metacarpals and forelimbs, 182
   paw, 25, 172
   pelvic girdle, 198
   radius, 132, 139
   scapula, 97
   tibia, 238
   ulna, 151, 160